Discrete Mathematical Structures with Applications to Combinatorics

Discrete Mathematical Structures with Applications to Combinatorics

V Ramaswamy
Professor and Head
Department of Information Science and Engineering
Bapuji Institute of Engineering and Technology
Davangare

Universities Press

Universities Press (India) Private Limited

Registered Office
3-5-747/1/A & 3-6-754/1 Himayatnagar
Hyderabad 500 029 (A.P.), India
e-mail: info@universitiespress.com

First published 2006
Reprinted 2008

ISBN-13: 978 81 7371 500 6
ISBN-10: 81 7371 500 9

Set in Times New Roman 10/12 by
Krishtel eMaging Solutions Pvt. Ltd., Chennai 600017.

Printed at
Sree Kalanjali Graphics
Hyderabad 500 029

Published by
Universities Press (India) Private Limited
3-5-747/1/A & 3-6-754/1 Himayatnagar
Hyderabad 500 029 (A.P.), India

Dedicated to the memory of my beloved mother, Smt. K. S. Alamelu, who was simple, innocent and the very embodiment of all good things

Contents

Preface

This book is meant to serve as an introduction to discrete mathematical structures. It covers most of the topics normally taught in a one-semester course at the undergraduate level. Students enrolled in BE programmes in computer science, information science and other engineering branches as well as in the MCA programme in Indian universities will find the book useful. A few chapters on combinatorics have been included to bring home the practical utility of the various concepts introduced in the book, and to enable appreciation of the myriad applications that apparently simple concepts such as relations, functions and the pigeon-hole principle have.

The book is organised as follows. Chapter 1 is concerned with the elements of mathematical logic, where both propositional calculus and predicate calculus are dealt with. Chapter 2 introduces sets and the various concepts associated with sets. Venn diagrams, operations that can be performed on sets, properties of these operations, De Morgan's laws, and the concept of partition, all find a place in this chapter.

Mathematical induction is a powerful tool for proving many results, and Chapter 3 is concerned with this principle. Chapter 4 deals with permutations and combinations, a topic of great interest to anyone who is interested in counting and combinatorial problems.

Algorithm analysis is usually done by formulating a recurrence relation and solving it. Recurrence relations are introduced at an elementary level in Chapter 5 and are dealt with in more detail in Chapter 12. This should be of particular interest to students of theoretical computer science.

Chapter 6 introduces the concept of relations and deals with topics such as equivalence relations, compatibility relation, manipulation of relations and Warshall's algorithm. Chapter 7 covers functions. How an innocent-looking concept like the pigeon-hole principle can be used to arrive at some profound results is also demonstrated in this chapter.

Boolean algebra plays a very important role in logic design and design of various electronic circuits. Finite boolean algebras are nothing but special lattices, and the two topics are covered in Chapter 8.

Graph theory, introduced in Chapter 9, is a topic of great interest not only to mathematicians and computer scientists but to electrical engineers as well. For instance, the electrical wiring in buildings can be modelled using graphs. Arriving at solutions to optimization problems such as the Travelling Salesman Problem using graph theory has been briefly covered in this chapter.

Chapters 10 and 11 deal with various algebraic structures like groups, rings and fields. In these two chapters, several results including Lagrange's theorem and the fundamental homomorphism theorem for groups have been proved. The capability for analytical thinking is a most essential trait in students of computer science, and the problems in these chapters are intended to facilitate the development of this skill.

In Chapter 13, the knowledge gained about functions is put to use in solving interesting combinatorial problems. Answers to questions such as 'How many functions from a finite set A to a finite set B are one–one and how many are onto?' are arrived at in this chapter. Chapter 14 is concerned with the concept of partition, which is nothing but the number of ways in which any positive integer can be written. The judicious ways in which one–one and onto functions and recurrence relations can be used in solving such problems need really to be appreciated.

Chapter 15 presents some miscellaneous applications of the concepts learned earlier in the book. The chapter offers solutions to eight challenging problems encountered in day-to-day life. Further, it discusses the concepts of derangement and rook polynomial. Also presented is the very interesting argument that if we want to count the number of elements in a finite set X, it is sufficient to count the number of elements in a related set Y, where the counting is easy to do, and then establish a one–one correspondence between the sets X and Y.

Acknowledgements

To begin, I wish to express my sincere gratitude to my father. But for his encouragement and persuasion, this book would not have become a reality. I owe a debt of gratitude to Shri Shamanur Shivashankarappa, the hon. secretary of the Bapuji Educational Association, Davangere, the late Professor G. S. Sivanna who was the director of the Bapuji Institute of Engineering and Technology, Davangere, its principal, Professor Y. Vrushabhendrappa, the chairman, Shri A. C. Jayanna, Shri Kasal S. Vittal and other management personnel for providing me an excellent and conducive atmosphere to work in. This book evolved out of the lectures I had delivered to the first batch of information science and engineering students at the Bapuji Institute of Engineering and Technology, and to a small group of computer science and engineering students whom I had taught combinatorics. My sincere thanks to them for being the testing ground for the material found in this book. I express my sincere thanks to the management of the University

College of Technology and Management, Malaysia (KUTPM). It was during my tenure there that the final proof readings were carried out. I am also very thankful to Ms H. B. Anitha and Mr K. S. Harish who helped me with the typing of parts of the manuscript. I express my sincere thanks to Universities Press for agreeing to publish this book. I am very grateful to the referees for their valuable suggestions. Finally, many thanks to my wife Ms S. Uma for supporting me with her infinite understanding and patience in the last few years when I was fully engrossed in the writing of the book.

I have referred to several books on the subject while writing this book. My thanks are due to all those authors and also to those who have directly or indirectly helped me in bringing out the book in its present form. This is my first humble attempt at writing a book, and all suggestions for its improvement are most welcome.

V. Ramaswamy

1 Elements of Mathematical Logic

1.1 Introduction and Motivation

Knowingly or unknowingly we make use of logic in our every day life. We make several kinds of statements which include interrogatives like "What is your name?", commands like "Do this work", statements like "Yesterday the maximum temperature was 38° C", which may be true or false. Mathematical logic is concerned with these kinds of statements and their truth values. This relatively old branch of mathematics is very much relevant even today. In this chapter we are mainly concerned with propositional calculus and predicate calculus. This chapter contains only very elementary notions of mathematical logic.

1.2 Propositional Calculus

In mathematical logic we are mainly concerned with sentences of the type "Yesterday I delivered a lecture in Belgaum" which may either be true or false. However, it may not be possible sometimes to determine whether a sentence is true or false. For example, the above sentence would be true if I did indeed deliver a lecture in Belgaum the previous day and false otherwise. Now consider the sentence "It may rain tomorrow". We cannot say this sentence is true. It is also not possible to say that this sentence is false. However, in mathematical logic we are concerned with sentences which are either true or false but not both. Such sentences are called statements or propositions. The values "true" and "false" are called the truth values of the proposition and are denoted by T and F respectively. Propositions, also called propositional variables, are denoted using small letters p, q, r, etc. They can stand for any proposition.

In our everyday life we combine several statements using "and" and "or" to obtain more complicated statements. For example, "Yesterday I went to Belgaum and delivered a lecture". In mathematical logic too we can combine various propositions using logical connectives to obtain more complicated propositions. The only difference between the two is that in our daily life when we combine sentences using "and" or "or", we expect the sentences to be related. Nobody will make a sentence like "Yesterday I went to Belgaum or I like music", whereas in mathematical logic we can combine totally unrelated statements using connectives.

Definitions

If p and q are two propositions, then "p and q" (denoted by $p \wedge q$) is a proposition whose truth value is T if and only if the truth values of both p and q are T. $p \wedge q$ is called conjunction of p and q.

"p or q" (denoted by $p \vee q$) is a proposition whose truth value is F if and only if the truth values of both p and q are F. $p \vee q$ is called disjunction of p and q.

$\sim p$ (also denoted by $\neg p$) is a proposition whose truth value is F if and only if the truth value of p is T. $\sim p$ is called the negation of p.

$p \rightarrow q$ is a proposition whose truth value is F if and only if the truth value of p is T and the truth value of q is F.

$p \leftrightarrow q$ is a proposition whose truth value is T if and only if the truth values of both p and q are T or the truth values of both p and q are F.

$p \rightarrow q$ is called a conditional statement or (it corresponds to the "if then" statement in English), while $p \leftrightarrow q$ is called a bi-conditional statement (it corresponds to "if and only if" statement in English).

When propositions are connected using connectives $\wedge, \vee, \sim, \rightarrow, \leftrightarrow$, we obtain compound propositions or compound statements. Each individual proposition which appears in the compound statement is called its component. Given the truth values of various components, the truth value of the compound statement can be obtained using a table called the truth table. This truth table will have a value for the compound statement for all possible values of the components that appear in the compound statement. Note that a compound statement can be formed using any number of propositions and connectives ($\wedge, \vee, \sim, \rightarrow, \leftrightarrow$) any number of times.

For two propositions p and q, the following truth table gives the values of $p \wedge q$, $p \vee q$, $\sim p$, $p \rightarrow q$ and $p \leftrightarrow q$.

p	q	$p \wedge q$	$p \vee q$	$p \rightarrow q$	$p \leftrightarrow q$
T	T	T	T	T	T
T	F	F	T	F	F
F	T	F	T	T	F
F	F	F	F	T	T

p	$\sim p$
T	F
F	T

The above truth table says that $p \to q$ can be F only when p is T and q is F. In all other cases $p \to q$ will be T. In other words, if p is T we expect q also to be T for $p \to q$ to be T. If p is F, we do not bother about what q is (T or F) – in both cases $p \to q$ is T.

Two compound propositions are said to be equivalent if the truth values of the compound propositions are identical for all possible combinations of truth values of various components which appear in the two propositions. If two compound propositions A and B are equivalent, we write $A \equiv B$. Consider the following truth table.

p	q	$p \to q$	$\sim p \vee q$	$q \to p$	$(p \to q) \wedge (q \to p)$	$p \leftrightarrow q$
T	T	T	T	T	T	T
T	F	F	F	T	F	F
F	T	T	T	F	F	F
T	T	T	T	T	T	T

Since the columns corresponding to $p \to q$ and $\sim p \vee q$ are identical, it follows that $p \to q \equiv \sim p \vee q$. We also note that $p \leftrightarrow q \equiv (p \to q) \wedge (q \to p)$.

The truth table of any compound statement involving two propositions will have four rows. The truth table of any compound statement involving three propositions will have 8 rows. In general, the truth table of any compound statement involving n propositions will have 2^n rows.

Definition

1. A compound proposition is said to be a tautology if its value always remains T whatever be the values of the various propositions that occur in the given proposition.
2. A compound proposition is said to be a contradiction if its value always remains F whatever be the values of the various propositions that occur in the given proposition.
3. A compound proposition is said to be satisfiable if its value is T for at least one combination of truth values of the propositions that occur in the given proposition.
4. A compound proposition is said to be a contingency if its value is T for some combination of truth values of the propositions that occur in the

given compound proposition, and F for some other combination of truth values of the propositions. In other words, if a compound proposition is a contingency, then its truth value can be both T and F.

The following are some well-known equivalences.

1.	$p \vee p \equiv p$	Idempotent property
2.	$p \wedge p \equiv p$	Idempotent property
3.	$p \vee q \equiv q \vee p$	Commutative property
4.	$p \wedge q \equiv q \wedge p$	Commutative property
5.	$p \vee (q \vee r) \equiv (p \vee q) \vee r$	Associative property
6.	$p \wedge (q \wedge r) \equiv (p \wedge q) \wedge r$	Associative property
7.	$p \vee (q \wedge r) \equiv (p \vee q) \wedge (p \vee r)$	Distributive property
8.	$p \wedge (q \vee r) \equiv (p \wedge q) \vee (p \wedge r)$	Distributive property
9.	$\sim (\sim p) \equiv p$	
10.	$\sim (p \vee q) \equiv \sim p \wedge \sim q$	De Morgan's laws
11.	$\sim (p \wedge q) \equiv \sim p \vee \sim q$	De Morgan's laws

Functionally complete set of connectives

From equivalences 9, 10 and 11 given above, we can see that

$$p \vee q \equiv \sim (\sim p \wedge \sim q) \quad \text{and} \quad p \wedge q \equiv \sim (\sim p \vee \sim q).$$

Thus $\vee$ can be expressed using only $\wedge$ and $\sim$. Similarly, $\wedge$ can be expressed using only $\vee$ and $\sim$. Thus it is not necessary to define both $\vee$ and $\wedge$. Suppose we have defined only $\wedge$ and $\sim$. Then, as we have seen earlier, we can obtain $\vee$. What about $\rightarrow$ and $\leftrightarrow$?

$p \rightarrow q \equiv \sim p \vee q \equiv \sim (p \wedge \sim q)$ says that $\rightarrow$ can be obtained using only $\wedge$ and $\sim$. Thus it is not necessary to define $\rightarrow$ also. Again,

$$p \leftrightarrow q \equiv (p \rightarrow q) \wedge (q \rightarrow p) \equiv [\sim (p \wedge \sim q)] \wedge [\sim (q \wedge \sim p)].$$

Hence $\leftrightarrow$ can also be obtained using only $\wedge$ and $\sim$, which means that it is not necessary to define $\leftrightarrow$. We can see that $\wedge$ cannot be expressed using only $\sim$. Also, $\sim$ cannot be expressed using only $\wedge$. But the other three connectives can be obtained using only $\wedge$ and $\sim$. We say that $\{\wedge, \sim\}$ is a functionally complete set of connectives. Similarly, we can show that $\{\vee, \sim\}$ is also a functionally complete set of connectives.

Example Construct a truth table for the compound proposition

$$((p \vee q) \wedge \sim (\sim p \wedge (\sim p \vee \sim r))) \vee (\sim p \wedge \sim q) \vee (\sim p \wedge \sim r).$$

Solution Since this compound proposition has 3 variables, the truth table will have $2^3 = 8$ rows.

p	q	r	$p \vee q$	$\sim p \vee \sim r$	$\sim p \wedge$ $(\sim p \vee \sim r)$	$\sim (\sim p \wedge$ $(\sim p \vee \sim r))$	$((p \vee q) \wedge$ $\sim (\sim p \wedge$ $(\sim p \vee \sim r)))$	$(\sim p \wedge \sim q)$	$(\sim p \wedge \sim r)$	Result
T	T	T	T	F	F	T	T	F	F	T
T	T	F	T	T	F	T	T	F	F	T
T	F	T	T	T	F	T	T	F	F	T
T	F	F	T	T	F	T	T	F	F	T
F	T	T	T	F	F	T	T	F	F	T
F	T	F	T	T	T	F	F	F	T	T
F	F	T	F	T	T	F	F	T	F	T
F	F	F	F	T	T	F	F	T	T	T

Since the last column is full of Ts, the given compound proposition is a tautology.

Note $p \vee (\sim p)$ is always T and hence it is a tautology whereas $p \wedge (\sim p)$ is always F and hence it is a contradiction.

If a compound statement contains n propositions, the number of rows in the truth table is 2^n which becomes unwieldy as n increases. Hence a truth table is not an efficient method of proving that a given compound statement is a tautology or that two compound statements are equivalent. We can use the properties mentioned above and simplify the given compound statement as the following example shows.

$$((p \vee q) \wedge \sim (\sim p \wedge (\sim q \vee \sim r))) \vee (\sim p \wedge \sim q) \vee (\sim p \wedge \sim r)$$

$$\equiv ((p \vee q) \wedge \sim ((\sim p \wedge \sim q) \vee (\sim p \wedge \sim r))) \vee (\sim p \wedge \sim q)$$
$$\vee (\sim p \wedge \sim r)$$

$$\equiv ((p \vee q) \wedge \sim (\sim p \wedge \sim q) \wedge \sim (\sim p \wedge \sim r)) \vee (\sim p \wedge \sim q)$$
$$\vee (\sim p \wedge \sim r)$$

$$\equiv ((p \vee q) \wedge (p \vee q) \wedge (p \vee r)) \vee (\sim p \wedge \sim q) \vee (\sim p \wedge \sim r)$$

$$\equiv ((p \vee q) \wedge (p \vee r)) \vee \sim (p \vee q) \vee \sim (p \vee r)$$

$$\equiv ((p \vee q) \wedge (p \vee r)) \vee \sim ((p \vee q) \wedge (p \vee r))$$

$$\equiv T.$$

Therefore $((p \vee q) \wedge \sim (\sim p \wedge (\sim q \vee \sim r))) \vee (\sim p \wedge \sim q) \vee (\sim p \wedge \sim r)$ is a tautology.

Definition The converse of $p \to q$ is defined to be the implication $q \to p$. The contrapositive of $p \to q$ is defined to be the implication $\sim q \to \sim p$.

If $p \to q$ is a tautology where p and q are compound statements, then we say that q logically follows from p. If $p_1 \wedge p_2 \wedge \cdots \wedge p_n \to q$ is a tautology, we say that q logically follows from $p_1, p_2, \ldots, p_n$. This means that if we know that all of $p_1, p_2, \ldots, p_n$ are true, then q must be true. Here $p_1, p_2, \ldots, p_n$ are called the premises or hypotheses and q is called the conclusion.

We know that if p is T and $p \to q$ is T, then q should be T. This can be stated as $p \wedge (p \to q) \Rightarrow q$. This rule is called modus ponens. From the above, it also follows that if $p \to q$ is T but q is F, then p should be F. For, if p is T then q should be T. This can be stated as $\sim q \wedge (p \to q) \Rightarrow \sim p$. This rule is called modus tollens.

Now consider the tautology $(p \to q) \leftrightarrow (\sim q \to \sim p)$. This result has the following application.

Suppose we assume p is T and we want to conclude q is T. For this we assume q is $F (\sim q$ is $T)$ and try to show that $\sim p$ is T (i.e., p is F). This method of concluding q from p is called the indirect method of proof.

Now consider the tautology $(p \to q) \wedge (\sim q) \to (\sim p)$. Suppose we want to prove q assuming p. We assume $\sim q$ and conclude $\sim p$. This method of proof is called proof by contradiction.

Suppose we want to prove by contradiction that q logically follows from $p_1, p_2, \ldots, p_n$. We assume q is false so that $\sim q$ is true and add it as an additional hypothesis. If we can show that $p_1 \wedge p_2 \wedge \cdots \wedge p_n \wedge \sim q$ leads to a contradiction, then at least one of $p_1, p_2, \ldots, p_n$ must be false. If $p_1, p_2, \ldots, p_n$ are all true, then $\sim q$ must be false, thus proving that q must be true.

Duality law

Consider a formula A consisting of propositional variables and the connectives $\wedge, \vee$ and $\sim$. A formula A' obtained from A by replacing $\wedge$ by $\vee$ and $\vee$ by $\wedge$ is said to be a dual of the formula A. Note that while taking dual, $\sim$ is not affected. If A contains T or F, then in the dual T is replaced by F and F is replaced by T.

Example Let A denote the formula $(p \wedge q \vee r) \wedge (p \wedge q \wedge r) \vee (\sim q \wedge \sim r \vee p)$. Then its dual is $(p \vee q \wedge r) \vee (p \vee q \vee r) \wedge (\sim q \vee \sim r \wedge p)$.

We can prove the following.

1. The negation of a formula is equivalent to its dual in which every variable is replaced by the negation of that variable.
2. If two formulas are equivalent, then so are their duals.

Well-formed formulas

A recursive definition is one where something is defined in terms of itself. Recursive definition of factorial n (denoted by $n!$) is as follows.

$$n! = 1 \text{ if } n = 0$$

$$= n \times (n-1)! \text{ if } n \geqslant 1.$$

Note that a factorial is defined in terms of the factorial. Hence this definition of factorial is a recursive definition. Note also that $n! = 1$ if $n = 0$ is a must. Otherwise, this definition will never terminate.

We will now give a recursive definition of well-formed formula (wff).

1. Any statement variable is a wff.
2. If A is a wff, then $\sim A$ is a wff.
3. If A and B are wffs, then $(A \wedge B)$, $(A \vee B)$, $(A \rightarrow B)$ and $(A \leftrightarrow B)$ are all wffs.
4. A string consisting of variables, connectives and parentheses is a wff if and only if it can be obtained by finitely many applications of the rules given above.

Rule 1 enables us to conclude that certain things are wffs. Rules 2 and 3 enable us to obtain more and more wffs. Rule 4 simply says that this is the only way to obtain wffs.

Some examples of wffs are $((p \wedge q) \vee r) \leftrightarrow s$, $p \vee (q \wedge s) \rightarrow t$, etc. Note that $\wedge$ is not a wff.

A formula A is said to be a substitution instance of another formula B if A can be obtained from B by substituting formulas for some variables in B taking care to substitute the same formula for the same variable whenever it occurs. For example, assume that B stands for the formula $r \rightarrow (t \wedge r)$. Substituting $a \leftrightarrow b$ for r in B, we obtain the formula A which is $a \leftrightarrow b \rightarrow (t \wedge (a \leftrightarrow b))$. Note that A is a substitution instance of B. However, $a \leftrightarrow b \rightarrow (t \wedge r)$ is not a substitution instance of B since the variable r in $t \wedge r$ has not been replaced by $a \leftrightarrow b$.

Normal forms

We will consider two normal forms, namely *disjunctive normal form* and *conjunctive normal form*. We will see that a given formula can have several disjunctive normal forms and several conjunctive normal forms. To obtain uniqueness, we will introduce the concepts of *principal disjunctive normal form* and *principal conjunctive normal form*.

We will use the word *sum* in place of disjunction and the word *product* in place of conjunction. Conjunction of variables and their negations is called an elementary product. Disjunction of variables and their negations is called an elementary sum. Part of an elementary product (sum), which itself is an elementary product (sum), is called a factor of the original elementary product (sum). For the elementary sum $p \vee q \vee \sim q$, some of the factors are $p \vee q, p \vee \sim q, \sim q$, etc.

A formula is said to be in disjunctive normal form if it is a sum of elementary products. Similarly, a formula is said to be in conjunctive normal form if it is a product of elementary sums.

Example Obtain a disjunctive normal form and a conjunctive normal form of the formula $p \wedge (p \rightarrow q)$.

We have

$$\begin{aligned} p \wedge (p \rightarrow q) &\equiv p \wedge (\sim p \vee q) \\ &\equiv (p \wedge \sim p) \vee (p \wedge q). \end{aligned}$$

Note that $(p \wedge \sim p)$ and $(p \wedge q)$ are elementary products. Hence $(p \wedge \sim p) \vee (p \wedge q)$ is a disjunctive normal form of $p \wedge (p \rightarrow q)$. Note that $p \wedge (\sim p \vee q)$ is a product of elementary sums. Hence it is a conjunctive normal form of $p \wedge (p \rightarrow q)$.

Consider the formula $p \vee (q \wedge r)$. Note that it is already in disjunctive normal form. We also have

$$\begin{aligned} p \vee (q \wedge r) &\equiv (p \vee q) \wedge (p \vee r) \\ &\equiv ((p \vee q) \wedge p) \vee ((p \vee q) \wedge r) \\ &\equiv (p \wedge p) \vee (q \wedge p) \vee (p \wedge r) \vee (q \wedge r). \end{aligned}$$

This is also a disjunctive normal form of $p \vee (q \wedge r)$. Hence the disjunctive normal form of a formula need not be unique.

Principal disjunctive normal form

Principal disjunctive normal form is also a disjunctive normal form in the sense that it is a disjunction of conjunctions. But in every conjunction, a variable or its negation but not both, will appear exactly once. We will agree to include either $p \wedge q$ or $q \wedge p$ but not both. For two variables p and q, we obtain $p \wedge q, p \wedge \sim q, q \wedge \sim p$ and $\sim p \wedge \sim q$ which are called min terms. The truth tables of these min terms are given below.

p	q	$p \wedge q$	$p \wedge \sim q$	$q \wedge \sim p$	$\sim p \wedge \sim q$
T	T	T	F	F	F
T	F	F	T	F	F
F	T	F	F	T	F
F	F	F	F	F	T

The last four columns show that no two of these min terms are equivalent. Moreover, each of them has the value T for exactly one combination of values of p and q (note that in the last four columns, in each row, T occurs exactly once). For a given formula, an equivalent formula consisting of disjunctions of min terms only is called its principal disjunctive normal form.

Principal conjunctive normal form

Principal conjunctive normal form is also a conjunctive normal form in the sense that it is a conjunction of disjunctions. But in every disjunction, a variable or its negation but not both, will appear exactly once. We will agree to include either $p \vee q$ or $q \vee p$, but not both. For two variables p and q, we obtain $p \vee q, p \vee \sim q, q \vee \sim p$ and $\sim p \vee \sim q$ which are called max terms. For a given formula, an equivalent formula consisting of conjunctions of max terms only is called its principal conjunctive normal form.

We will now illustrate how to obtain principal disjunctive normal form and principal conjunctive normal form of a given formula.

Example The truth table of a formula A is given below. Obtain its principal disjunctive and principal conjunctive normal forms.

p	q	r	A
T	T	T	F
T	T	F	F
T	F	T	T
T	F	F	F
F	T	T	T
F	T	F	T
F	F	T	F
F	F	F	T

To obtain principal disjunctive normal form, we first look at the rows corresponding to which A has the value T. A has the value T corresponding to $p = T, q = F$ and $r = T$. Hence we consider the min term $p \wedge \sim q \wedge r$. (Note that when a variable has truth value T, we take that variable and when a variable has truth value F, we take the negation of that variable). The next T occurs corresponding to $p = F, q = T$ and $r = T$. Hence we consider the min term $\sim p \wedge q \wedge r$. Proceeding like this we obtain two more min terms $\sim p \wedge q \wedge \sim r$ and $\sim p \wedge \sim q \wedge \sim r$. We thus obtain the principal disjunctive normal form $(p \wedge \sim q \wedge r) \vee (\sim p \wedge q \wedge r) \vee (\sim p \wedge q \wedge \sim r) \vee (\sim p \wedge \sim q \wedge \sim r)$.

To obtain principal conjunctive normal form, we first look at the rows corresponding to which A has the value F. A has the value F corresponding to $p = T, q = T$ and $r = T$. Hence we consider the max term $\sim p \vee \sim q \vee \sim r$. (Note here that when a variable has truth value F, we take that variable and when a variable has truth value T, we take the negation of that variable). Next F occurs corresponding to $p = T, q = T$ and $r = F$. Hence we consider the max term $\sim p \vee \sim q \vee r$. Proceeding like this we obtain two more max terms $\sim p \vee q \vee r$ and $p \vee q \vee \sim r$. We thus obtain the principal conjunctive normal form $(\sim p \vee \sim q \vee \sim r) \wedge (\sim p \vee \sim q \vee r) \wedge (\sim p \vee q \vee r) \wedge (p \vee q \vee \sim r)$.

Using the truth table to determine the principal disjunctive and the principal conjunctive normal forms is not a desirable method. This is due to the fact that if a formula contains n propositional variables, then the truth table will have 2^n rows. The number of rows increases enormously as n increases and this will make the construction of the truth table a very difficult exercise.

The following example will show that the truth table is not the only method to obtain principal disjunctive and principal conjunctive normal forms of a given formula. To obtain principal disjunctive normal form, we will make use of the fact that any formula A is equivalent to $A \wedge T$ (if A is T, then $A \wedge T$ is T while if A is F, then $A \wedge T$ is F), which is again equivalent to $A \wedge (p \vee \sim p)$ for any propositional variable p. Note that $p \vee \sim p$ is always a tautology. To obtain the principal conjunctive normal form, we will make use of the fact that any formula A is equivalent to $A \vee F$ (if A is T, then $A \vee F$ is T while if A is F, then $A \vee F$ is F) which is again equivalent to $A \vee (q \wedge \sim q)$ for any propositional variable q. Note that $q \wedge \sim q$ is always a contradiction.

If we know the principal disjunctive normal form of a formula A, we can get its principal conjunctive normal form as follows.

The principal disjunctive normal form of $\sim A$ is the disjunction of the remaining min terms which do not appear in the principal disjunctive normal form of A. We will then take $\sim\sim A$ which is nothing but A. This will give the principal conjunctive normal form of A. Similarly, from the principal conjunctive normal form of A, we can obtain its principal disjunctive normal form. This method is also illustrated in the following example.

Example Obtain the principal disjunctive and principal conjunctive normal forms of the formula A given by $(\sim p \rightarrow r) \wedge (q \leftrightarrow p)$.

Solution We will first determine the principal conjunctive normal form, and using it determine the principal disjunctive normal form. We have

$$\begin{aligned}(\sim p \to r) \wedge (q \leftrightarrow p) &\equiv (p \vee r) \wedge ((q \to p) \wedge (p \to q))\\
&\equiv (p \vee r) \wedge (\sim q \vee p) \wedge (\sim p \vee q)\\
&\equiv ((p \vee r) \vee (q \wedge \sim q)) \wedge ((\sim q \vee p)\\
&\quad \vee (r \wedge \sim r)) \wedge ((\sim p \vee q) \vee (r \wedge \sim r))\\
&\equiv (p \vee r \vee q) \wedge (p \vee r \vee \sim q) \wedge (\sim q \vee p \vee r)\\
&\quad \wedge (\sim q \vee p \vee \sim r) \wedge (\sim p \vee q \vee r)\\
&\quad \wedge (\sim p \vee q \vee \sim r)\\
&\equiv (p \vee q \vee r) \wedge (p \vee \sim q \vee r)\\
&\quad \wedge (p \vee \sim q \vee \sim r) \wedge (\sim p \vee q \vee r)\\
&\quad \wedge (\sim p \vee q \vee \sim r).\end{aligned}$$

This is the principal conjunctive normal form. What are the missing max terms? The remaining three (note that there are totally $2^3 = 8$ max terms out of which 5 have appeared above) are $p \vee q \vee \sim r$, $\sim p \vee \sim q \vee r$ and $\sim p \vee \sim q \vee \sim r$. Their conjunction will give $\sim A$ so that

$$\sim A \equiv (p \vee q \vee \sim r) \wedge (\sim p \vee \sim q \vee r) \wedge (\sim p \vee \sim q \vee \sim r).$$

Taking negation on both sides and using De Morgan's laws, we obtain

$$\begin{aligned}A &\equiv \sim\sim A \equiv \sim((p \vee q \vee \sim r) \wedge (\sim p \vee \sim q \vee r)\\
&\quad \wedge (\sim p \vee \sim q \vee \sim r)).\\
&\equiv \sim(p \vee q \vee \sim r) \vee \sim(\sim p \vee \sim q \vee r) \vee \sim(\sim p \vee \sim q \vee \sim r)\\
&\equiv (\sim p \wedge \sim q \wedge r) \vee (p \wedge q \wedge \sim r) \vee (p \wedge q \wedge r).\end{aligned}$$

Note that the above is the principal disjunctive normal form of A.

Principal disjunctive normal form can also be used to prove the equivalence of formulae as is demonstrated in the next example.

Example Prove that $p \vee (\sim p \wedge q)$ and $p \vee q$ are equivalent.

Solution We will show that the principal disjunctive normal form of both are the same. This will prove that the formulae are equivalent. We have

$$\begin{aligned}p \vee (\sim p \wedge q) &\equiv (p \wedge (q \vee \sim q)) \vee (\sim p \wedge q)\\
&\equiv (p \wedge q) \vee (p \wedge \sim q) \vee (\sim p \wedge q).\end{aligned}$$

$$\begin{aligned}p \vee q &\equiv (p \wedge (q \vee \sim q)) \vee (q \wedge (p \vee \sim p))\\
&\equiv (p \wedge q) \vee (p \wedge \sim q) \vee (q \wedge p) \vee (q \wedge \sim p)\\
&\equiv (p \wedge q) \vee (p \wedge \sim q) \vee (\sim p \wedge q).\end{aligned}$$

Theory of inference for propositional calculus

The main function of logic is to provide rules of inference or principles of reasoning. The theory associated with these rules is referred to as inference theory. Here we are concerned with determining whether a conclusion can follow from a given set of premises. One premise may be "If I go to Chennai, I will meet my friend". Another premise may be "I go to Chennai". The conclusion may be "I will meet my friend". We can intuitively see that the conclusion follows from the given premises. Suppose the premises are "If I go to Chennai, I will meet my friend" and "I will meet my friend". Then the conclusion "I go to Chennai" obviously does not follow from the given premises. Note that the premises have to be taken as facts whether they make sense or not from the view point of the world. In the above premises, if the word "Chennai" is replaced by the word "heaven", we will have to accept the premises as given facts even though "I go to heaven" does not make any sense in the real world.

Making use of the given set of premises and the accepted rules of reasoning, we determine whether the given conclusion follows or not. This method of derivation is called deduction or formal proof. Any conclusion which is arrived at by following the rules of inference is called a valid conclusion and the argument used is called a valid argument.

Suppose P is a premise and C is a conclusion. We say that "C logically follows from P" or "C is a valid conclusion of the premise P" if and only if $P \to C$ is a tautology. We denote this by writing $P \Rightarrow C$.

Instead of just one premise P, suppose we have a set of m premises $H_1, H_2, \ldots, H_m$. We then say that "C logically follows from these m premises" or "C is a valid conclusion of these m premises" if and only if $H_1 \wedge H_2 \wedge \cdots \wedge H_m \Rightarrow C$.

We will now see how to determine whether C logically follows from $H_1, H_2, \ldots, H_m$ or not. Let $p_1, p_2, \ldots, p_n$ denote all the atomic variables which occur in m premises and conclusion C. For all possible combinations of truth values of $p_1, p_2, \ldots, p_n$, determine the truth values of the m premises and the conclusion C. Look at any row corresponding to which all of $H_1, H_2, \ldots, H_m$ are true. For this row, the value of C must be true. Equivalently, corresponding to any row in which C is F, at least one of $H_1, H_2, \ldots, H_m$ must be F.

We will make use of the following rules of inference.

Rule A: A premise can be introduced at any point in the derivation.

Rule B: A formula S can be introduced in a derivation if S is tautologically implied by one or more of the preceding formulas in the derivation.

Rule CP: If we can derive S from R and a set of premises, then we can derive $R \to S$ from the set of premises alone.

Consistency of premises and indirect method of proof

Consider m premises $H_1, H_2, \ldots, H_m$. Let $p_1, p_2, \ldots, p_n$ denote all atomic variables which occur in these m premises. These premises $H_1, H_2, \ldots, H_m$ are said to be consistent if we can assign truth values to $p_1, p_2, \ldots, p_n$ corresponding to which all of $H_1, H_2, \ldots, H_m$ are true. Note that this is equivalent to saying that corresponding to these values of $p_1, p_2, \ldots, p_n$, $H_1 \wedge H_2 \wedge \cdots \wedge H_m$ is T. The premises are said to be inconsistent if for all possible values of $p_1, p_2, \ldots, p_n$, $H_1 \wedge H_2 \wedge \cdots \wedge H_m$ is F. This means at least one of the premises $H_1, H_2, \ldots, H_m$ is F.

We will now see how the consistency of premises and deriving that a conclusion C follows from a set of premises are related.

Again, assume that we have a set of premises $H_1, H_2, \ldots, H_m$ and we want to show that a conclusion C follows from these premises. We will use indirect method of proof, also called the method of contradiction. As the name implies, we will assume that C is false so that $\sim C$ can be regarded as an additional premise. We will then show that the $m + 1$ premises $H_1, H_2, \ldots, H_m, \sim C$ are inconsistent. This will mean that $H_1 \wedge H_2 \wedge \cdots \wedge H_m \wedge \sim C$ is F. Hence if $H_1 \wedge H_2 \wedge \cdots \wedge H_m$ is T, then $\sim C$ must have a truth value F so that the truth value of C is T.

We will now illustrate the above concepts by means of examples.

Solved problems

1. Show that the premises $p \to q$, $p \to r$, $q \to \sim r$ and p are inconsistent.

Solution

{1}	(1)	$p \to q$	Rule A
{2}	(2)	$q \to \sim r$	Rule A
{1, 2}	(3)	$p \to \sim r$	T, (1), (2)
{4}	(4)	p	Rule A
{1, 2, 4}	(5)	$\sim r$	T, (3), (4)
{6}	(6)	$p \to r$	Rule A
{4, 6}	(7)	r	T, (4), (6)
{1, 2, 4, 6}	(8)	$r \wedge \sim r$	T, (5), (7)

Note that $r \wedge \sim r$ is a contradiction which proves that the premises are inconsistent.

The numbers within parentheses denote statement numbers whereas the numbers within curl brackets indicate the numbers of those statements which are used in deriving the statements mentioned against them. For example, consider

$\{1, 2, 4\}$ (5) $\sim r$ T, (3), (4)

Here 5 is the number of this statement. We derive $\sim r$ using statement 3, namely $p \rightarrow\sim r$ and statement 4, namely p. But how do we get $p \rightarrow\sim r$? Using statements 1 and 2 ($p \rightarrow q$ and $q \rightarrow\sim r$). That is why the numbers 1, 2 and 4 that are responsible for deriving $\sim r$ are included in curl brackets. Note that we have not included 3 because 3 itself follows from 1 and 2.

2. Prove that D follows from $A \rightarrow B$, $A \rightarrow C$, $\sim (B \wedge C)$ and $D \vee A$.

Solution We will make use of indirect method of proof. We will assume $\sim D$ and obtain a contradiction.

$\{1\}$	(1)	$\sim (B \wedge C)$	Rule A
$\{1\}$	(2)	$\sim B \vee \sim C$	T, (1)
$\{1\}$	(3)	$B \rightarrow\sim C$	T, (2)
$\{4\}$	(4)	$D \vee A$	Rule A
$\{4\}$	(5)	$\sim D \rightarrow A$	T, (4)
$\{6\}$	(6)	$A \rightarrow B$	Rule A
$\{1, 6\}$	(7)	$A \rightarrow\sim C$	T, (3), (6)
$\{1, 4, 6\}$	(8)	$\sim D \rightarrow\sim C$	T, (5), (7)
$\{9\}$	(9)	$\sim D$	Assumed premises
$\{1, 4, 6, 9\}$	(10)	$\sim C$	T, (8), (9)
$\{11\}$	(11)	$A \rightarrow C$	Rule A
$\{4, 11\}$	(12)	$\sim D \rightarrow C$	T, (5), (11)
$\{13\}$	(13)	$\sim D$	Assumed premises
$\{4, 11, 13\}$	(14)	C	T, (12), (13)
$\{1, 4, 6, 9, 11, 13\}$	(15)	$C \wedge \sim C$	T, (14), (10)

3. Prove that $A \rightarrow (B \rightarrow C)$, $D \rightarrow (B \wedge \sim C)$ and $A \wedge D$ are inconsistent.

Solution

{1}	(1)	$A \wedge D$	Rule A
{1}	(2)	A	T, (1)
{1}	(3)	D	T, (1)
{4}	(4)	$A \to (B \to C)$	Rule A
{1, 4}	(5)	$B \to C$	T, (2), (4)
{1, 4}	(6)	$\sim B \vee C$	T, (5)
{7}	(7)	$D \to (B \wedge \sim C)$	Rule A
{1, 7}	(8)	$B \wedge \sim C$	T, (3), (7)
{1, 7}	(9)	$\sim (\sim B \vee C)$	T, (8)
{1, 4, 7}	(10)	$(\sim B \vee C) \wedge \sim (\sim B \vee C)$	T, (6), (9)

4. Prove the following.

$$(R \to \sim Q), R \vee S, S \to \sim Q, P \to Q \Rightarrow \sim P.$$

Solution We will assume P and obtain a contradiction.

{1}	(1)	$P \to Q$	Rule A
{2}	(2)	P	Assumed premise
{1, 2}	(3)	Q	T, (1), (2)
{4}	(4)	$S \to \sim Q$	Rule A
{1, 2, 4}	(5)	$\sim S$	T, (3), (4)
{6}	(6)	$R \vee S$	Rule A
{6}	(7)	$\sim R \to S$	T, (6)
{8}	(8)	$R \to \sim Q$	Rule A
{1, 2, 8}	(9)	$\sim R$	T, (3), (8)
{1, 6, 2, 8}	(10)	S	T, (7), (9)
{1, 2, 4, 6, 8}	(11)	$S \wedge \sim S$	T, (5), (10)

Exercises

1. Which of the following are statements?

   (i) Is 2.7 an integer?
   (ii) Meet me tomorrow.
   (iii) In India it is very cold in May.
   (iv) If the finance company closes, then we will be in trouble.

2. Determine the conjunction and disjunction of p and q where

$$p : 3 + 1 > 5 \quad q : 12 = 4 \times 3.$$

3. Determine the truth value of each of the following statements.

   (i) $2 > 3$ and -38 is a negative integer.
   (ii) -8 is not a natural number.
   (iii) $-14 \leq -29$ or $64 = 23 \times 2$.

4. Which of the following statements is the negative of the statement "58 is even or -17 is negative"?

   (i) 58 is even or -17 is not negative.
   (ii) 58 is odd or -17 is not negative.
   (iii) 58 is even and -17 is not negative.
   (iv) 58 is odd and -17 is not negative.

5. Let p stand for the statement "Today is Friday", q stand for the statement "I will watch a movie" and r stand for the statement "$3 + 2$ is greater than 7". Write an English sentence which corresponds to each of the following.

   (i) $\sim r \wedge q$
   (ii) $\sim q \vee p$
   (iii) $\sim (p \vee r)$
   (iv) $p \vee \sim r$

6. Consider p, q and r given in the above problem. Write each of the following in terms of p, q, r and using connectives.

   (i) Today is Friday and $3 + 2$ is not greater than 7.
   (ii) Either I will watch a movie or today is Friday.
   (iii) Today is not Friday and I will not watch a movie.

7. Suppose we define $p \downarrow q$ as follows.

p	q	$p \downarrow q$
T	T	F
T	F	F
F	T	F
F	F	T

Construct a truth table for each of the following.

(i) $(p \downarrow q) \downarrow r$
(ii) $(p \downarrow q) \wedge (p \downarrow r)$
(iii) $(p \downarrow q) \downarrow (p \downarrow r)$

Solution for (ii) above

p	q	r	$p \downarrow q$	$p \downarrow r$	$(p \downarrow q) \wedge (p \downarrow r)$
T	T	T	F	F	F
T	T	F	F	F	F
T	F	T	F	F	F
F	T	T	F	F	F
F	F	T	T	F	F
F	T	F	F	T	F
T	F	F	F	F	F
F	F	F	T	T	T

8. Suppose p stands for the statement "I am healthy", q stands for the statement "I always walk" and r stands for the statement "I want to buy a car". Write each of the following statements in terms of p, q, and r using connectives.

   (i) I am healthy implies that I always walk.
   (ii) I want to buy a car if I am healthy.
   (iii) My always walking is sufficient for me to be healthy.
   (iv) Being healthy is necessary for me to buy a car.

9. State the converse of each of the following.

   (i) If $8 \times 9 = 72$, then I will not attend class.
   (ii) If I am early, then I would have come by car.
   (iii) If my work is over in time and I have the mood, then I will go for a movie.

10. Determine the truth value of each of the following.

    (i) If 7 is odd, then New Delhi is the capital of India.
    (ii) If 12 is even, then Chennai is the capital of India.
    (iii) If 31 is even, then New Delhi is the capital of India.
    (iv) If 75 is even, then Chennai is the capital of India.

11. Let p denote the statement "I will join the civil services"; q denote the statement "I will buy a house"; and r denote the statement "I like music". Write English sentences corresponding to the following statements.

(i) $\sim p \wedge q \rightarrow r$
(ii) $r \rightarrow p \vee q$
(iii) $\sim r \rightarrow \sim q \vee p$
(iv) $q \wedge \sim q \leftrightarrow r$

12. Construct truth tables to determine whether the following is a tautology, contradiction or contingency.

(i) $p \rightarrow (q \rightarrow p)$
(ii) $q \rightarrow (q \rightarrow p)$
(iii) $q \wedge (\sim q \wedge p)$
(iv) $(q \wedge p) \vee (q \wedge \sim p)$
(v) $(p \wedge q) \rightarrow p$
(vi) $p \rightarrow (q \wedge p)$

13. Given that $p \rightarrow q$ is F, determine the truth value of $(\sim (p \wedge q)) \rightarrow q$.
14. Show that $(p \downarrow p) \downarrow (q \downarrow q) \equiv p \wedge q$.
15. To describe the various restaurants in the city, we let p denote the statement, "The food is good", q denote the statement, "The service is good", and r denote the statement "The rating is three-star". Write the following statements in symbolic form.
(a) Either the food is good or the service is good or both.
(b) Either the food is good or the service is good but not both.
(c) The food is good while the service is poor.
(d) It is not the case that both the food is good and the rating is three-star.
(e) If both the food and services are good, then the rating will be three-star.
(f) It is not true that a three-star rating always means good food and good service.

16. Let p denote the statement "The material is interesting" and q denote the statement "The exercises are challenging", and r denote the statement "The course is enjoyable". Write the following statements in symbolic form.

(a) The material is interesting and the exercises are challenging.
(b) The material is uninteresting, the exercises are not challenging and the course is not enjoyable.
(c) If the material is not_interesting and the exercises are not challenging, then the course is not enjoyable.
(d) The material is interesting means the exercises are challenging and conversely.
(e) Either the material is interesting or the exercises are not challenging but not both.

17. Write the following statements in symbolic form.

    (a) The sun is bright and the humidity is not high.
    (b) If I finish my homework before dinner and it does not rain, then I will go to see a movie.
    (c) If you do not see me tomorrow, it means I have gone to Chennai.
    (d) If the utility cost goes up or the request for additional funding is denied, then a new computer will be purchased if and only if we can show that the current computing facilities are indeed not adequate.

18. A certain country is inhabited only by people who either always tell the truth or always tell lies and who will respond to questions only with a "yes" or a "no". A tourist comes to a fork in the roac where one branch leads to the capital and the other does not. There is no sign indicating which branch to take but there is an inhabitant Mr. Z standing at the fork. What single question should the tourist ask him to determine which branch to take ?

19. Obtain the principal conjunctive normal form of the following formulae.

    (a) $(p \wedge q \wedge r) \vee (\sim p \wedge q \wedge r) \vee (\sim p \wedge \sim q \wedge \sim r)$
    (b) $(p \wedge q) \vee (\sim p \wedge q) \vee (p \wedge \sim q)$
    (c) $(p \wedge q) \vee (\sim p \wedge q \wedge r)$

20. Obtain the principal disjunctive and the principal conjunctive normal forms of the following formulae.

    (i) $q \wedge (p \vee \sim q)$
    (ii) $p \vee (\sim p \rightarrow (q \vee (\sim q \rightarrow r)))$
    (iii) $(p \rightarrow (q \wedge r)) \wedge (\sim p \rightarrow (\sim q \wedge \sim r))$
    (iv) $p \rightarrow (p \wedge (q \rightarrow p))$
    (v) $(q \rightarrow p) \wedge (\sim p \wedge q)$

21. Derive the following.

    (i) $\sim p \vee q, \sim q \vee r, r \rightarrow s \Rightarrow p \rightarrow s$
    (ii) $p, p \rightarrow (q \rightarrow (r \wedge s)) \Rightarrow q \rightarrow s$
    (iii) $p \rightarrow q \Rightarrow p \rightarrow (p \wedge q)$
    (iv) $(p \vee q) \rightarrow r \Rightarrow (p \wedge q) \rightarrow r$
    (v) $p \rightarrow (q \rightarrow r), q \rightarrow (r \rightarrow s) \Rightarrow p \rightarrow (q \rightarrow s)$
    (vi) $(r \rightarrow \sim q), r \vee s, s \rightarrow \sim q, p \rightarrow q \Rightarrow \sim p$

1.3 Predicate Calculus

Propositional calculus which we have seen so far, though powerful, is not helpful in finding out the relationships that may exist between statements. For example, consider the statements "Lion is an animal" and "Tiger is an animal". We can represent the first statement using a propositional variable, say p, and the second statement using a propositional variable, say q. We know that the above two statements are not totally unrelated since both of them say that something is an animal. The variables p and q do not convey this message. Suppose we represent the above statements as animal (Lion), animal (Tiger). Now the relationship between the two statements, namely both Lion and Tiger are animals, has been brought out clearly. Here animal is called a predicate and Lion the argument for the predicate. Tiger is another argument.

A predicate is something which is either true or false. For example, animal(x) is true if x is indeed an animal. For example, animal(cow) and animal(dog) are true whereas animal(lemon) is false. Animal is called a one-place predicate since it has only one argument. In general, a predicate can have any number of arguments. Consider the predicate le which stands for "less than or equal to". This predicate needs two arguments: $le(a, b)$ is true if and only if a is less than or equal to b. For example, $le(5, 5)$ is true whereas $le(8, 3)$ is false.

Many times we may want to talk about a property which is possessed by several members in a group. For example, consider the statement "All men eat rice". Such statements cannot be comfortably represented using propositional calculus and one has to go for predicate calculus to represent these statements efficiently. The above statement can be interpreted as: for all x, if x is a man, then x eats rice. Here we have one predicate "man" with one argument and another predicate "eats" which has two arguments. Now, the above statement in predicate calculus is represented as $(\forall x)$ man $(x) \rightarrow$ eats $(x,$ rice$)$. Here $(\forall x)$ (also represented as (x)) is called a *universal quantifier*.

Now consider the statement "Some animals are dangerous". This statement can be interpreted as "There is at least one animal which is dangerous". In predicate calculus this statement can be represented as $(\exists x)$ animal $(x) \wedge$ dangerous (x). Here animal and dangerous are both one-place predicates. $(\exists x)$ is called an *existential quantifier*.

The following equivalences hold good.

1. $\sim ((\forall x)p(x)) \equiv (\exists x) \sim p(x)$.
   The negation of the statement "All men eat rice" is obviously "There is at least one man who does not eat rice".

2. $\sim ((\exists x) \sim p(x)) \equiv (\forall x)p(x)$.
(If $p(x)$ stands for "x has property p", then $(\exists x) \sim p(x)$ stands for "There is something which does not have property p". Its negation is obviously "everything has property p"). The negation of the statement "There is an animal which is not dangerous" is "All animals are dangerous".
3. $(\exists x)(p(x) \to q(x)) \equiv (\forall x)p(x) \to (\exists x)q(x)$.
4. $(\exists x)(p(x) \vee q(x)) \equiv (\exists x)p(x) \vee (\exists x)q(x)$.
5. $(\forall x)(p(x) \wedge q(x)) \equiv (\forall x)p(x) \wedge (\forall x)q(x)$.

Free and bound variables

Suppose a formula contains a part of the form $(x)P(x)$ or $(\exists x)P(x)$. We will call such a part an x-bound part of the formula. Any occurrence of x in an x-bound part of a formula is called a *bound occurrence* of x. We will call any occurrence of x that is not a bound occurrence a free occurrence. $P(x)$ is referred to as the scope of the quantifier. Thus the scope of a quantifier is nothing but the formula which immediately follows the quantifier.

Example Consider the following.

1. $(x)P(x, y)$
2. $(x)(P(x) \to Q(x))$
3. $(x)(P(x) \to (\exists y)Q(x, y))$
4. $(x)(P(x) \to Q(x)) \vee (x)(P(x) \to R(x))$
5. $(\exists x)(P(x) \wedge Q(x))$
6. $(\exists x)P(x) \wedge Q(x)$

In (1), $P(x, y)$ is the scope of the quantifier. Both occurrences of x are bound occurrences whereas the occurrence of y is a free occurrence. In (2) the scope of the universal quantifier is $P(x) \to Q(x)$ and all occurrences of x are bound occurrences. In (3) the scope of the universal quantifier is $P(x) \to (\exists y)Q(x, y)$. The scope of the existential quantifier is $Q(x, y)$. All occurrences of x and y are bound occurrences. In (4) the scope of the first quantifier is $P(x) \to Q(x)$ and the scope of the second quantifier is $P(x) \to R(x)$. All occurrences of x are bound occurrences. In (5) the scope of the existential quantifier is $P(x) \wedge Q(x)$, whereas in (6) the scope of the existential quantifier is $P(x)$. Occurrence of x in $Q(x)$ is a free occurrence.

In the bound occurrence of a variable, the letter which is used to represent the variable is not at all important. We can use any other letter (which is not used anywhere else in the formula) without affecting the formula. For example, the formulas $(x)P(x, y)$ and $(v)P(v, y)$ are the same. Note that the bound occurrence of a variable cannot be replaced by a constant, whereas the free occurrence of a variable can be replaced by a constant. $(x)P(x) \wedge Q(d)$ is a substitution instance of $(x)P(x) \wedge Q(z)$ and is expressed in English as: every x has the property P and d has the property Q. Note that in (6), x which occurs in P and x which occurs in Q are not the same. x which occurs in P is a bound occurrence whereas x which occurs in Q is a free occurrence. We can distinguish the two by using two different letters. For example, (6) can be written as $(\exists y)P(y) \wedge Q(x)$.

Universe of discourse

We will assume that all the variables which are quantified stand for only those objects which are members of a particular set. This restricted class is referred to as the universe of discourse. If the discussion refers to only human beings, then the universe of discourse will be the class of all human beings.

We will now look at the advantage of this restriction. Consider the statement "All human beings are mortal". If $H(x)$ stands for "x is a human being" and $M(x)$ stands for "x is a mortal", then the above statement can be written as $(x)(H(x) \rightarrow M(x))$. If our universe of discourse is the class of all human beings, then the statement can simply be represented as $(x)M(x)$.

As another example, consider the statement "For every positive integer there is a larger positive integer". If $P(x)$ stands for "x is a positive integer" and $G(u, v)$ stands for "u is greater than v", then the above statement can be represented as $(x)(P(x) \rightarrow (\exists y)(P(y) \wedge G(y, x)))$. If our universe of discourse is the set of all positive integers, then the above statement can be simply represented as $(x)(\exists y)G(y, x)$.

The truth value of a statement depends on the universe of discourse. For example, consider the predicate $L(x)$ which stands for "x is less than 8" and the statements $(x)L(x)$ and $(\exists x)L(x)$. Consider the following universe of discourse.

(i) $\{6, 5, 2, 1\}$
(ii) $\{4, 11, 5\}$
(iii) $\{16, 43, 9\}$

$(x)L(x)$ is true for (i) but it is false for (ii) and (iii). $(\exists x)L(x)$ is true for (i) and (ii) but it is false for (iii).

Inference theory for predicate calculus

A major difference between inference theory for propositional calculus and inference theory for predicate calculus lies in the fact that the latter uses quantifiers (universal and existential) which are not available for the former. Hence inference theory for predicate calculus makes use of the rules of inference available for propositional calculus. We also require additional rules for dealing with formulas involving quantifiers. For eliminating quantifiers during the course of derivation, we will use two rules of specifications, viz., rules US and ES. Once the quantifiers are eliminated, the derivation proceeds exactly as in the case of propositional calculus and we arrive at the conclusion. If the conclusion is to be quantified, we again make use of two rules of generalizations, viz. rules UG and EG.

The major task for us is to specify the rules US, ES, UG and EG. Let $A(x)$ denote a formula where the occurrence of x is free and $A(y)$ denote a formula which is obtained by the substitution of y for x in $A(x)$. We know that for such a substitution $A(x)$ must be free for y.

Rule US: From $(x)A(x)$, we can conclude $A(y)$.

Naturally, if $A(x)$ is true for all x, then $A(y)$ must be true.

Rule ES: From $(\exists x)A(x)$ we can conclude $A(y)$ provided y is not free in any given premise and also not free in any prior step of the derivation. We can meet these requirements by choosing a new variable every time ES is used.

Rule EG: From $A(x)$ we can conclude $(\exists y)A(y)$.

Rule UG: From $A(x)$ we can conclude $(y)A(y)$ provided x is not free in any of the given premises and provided that if x is free in a prior step which resulted from the use of ES, then no variables introduced by that use of ES appear free in $A(x)$.

We will now justify the restriction in rule UG by an example.

Example Assume that $M(u, v)$ stands for the fact that "u is a multiple of v". Suppose the universe of discourse is $\{8, 12, 14, 15\}$. Then the statement $(\exists x)M(x, 3)$ is true because $M(12, 3)$ and $M(15, 3)$ are true. But $(y)M(y, 3)$ is false since $M(8, 3)$ and $M(14, 3)$ are not true. We consider the following derivation. What is wrong with it?

$\{1\}$	$(1)(\exists u)M(u, 3)$	Rule A
$\{1\}$	$(2)M(x, 3)$	ES, (1)
$\{1\}$	$(3)(y)M(y, 3)$	UG without second restriction, (2)

Note that from $M(x, 3)$ we have drawn the conclusion $(y)M(y, 3)$ in step 3. x is not free in the premise and hence the first restriction in UG is satisfied. But x is free in step 2 which resulted by the use of ES. Also, x has been introduced by the use of ES and appears free in $M(x, 3)$. Hence it cannot be generalised.

Solved problems

1. Prove that the conclusion "Sita is a mortal" follows from the premises "All human beings are mortal" and "Sita is a human being".

   Solution Let $H(x)$ denote "x is a human being" and $M(x)$ denote "x is a mortal" and let s denote Sita. Then we have

{1}	(1)	$(x)(H(x) \to M(x))$	Rule A
{1}	(2)	$H(s) \to M(s)$	US, (1)
{3}	(3)	$H(s)$	Rule A
{1, 3}	(4)	$M(s)$	T, (2), (3).

2. Show that $(x)((P(x) \to Q(x)) \wedge (Q(x) \to R(x))) \Rightarrow (x)(P(x) \to R(x))$.

   Solution

{1}	(1)	$(x)(P(x) \to Q(x))$	Rule A
{1}	(2)	$P(y) \to Q(y)$	US, (1)
{3}	(3)	$(x)(Q(x) \to R(x))$	Rule A
{3}	(4)	$Q(y) \to R(y)$	US, (3)
{1, 3}	(5)	$P(y) \to R(y)$	T, (2), (4)
{1, 3}	(6)	$(x)(P(x) \to R(x))$	UG, (5)

3. Prove that $(\exists x)M(x)$ follows logically from the premises $(x)(H(x) \to M(x))$ and $(\exists x)H(x)$.

   Solution

{1}	(1)	$(\exists x)H(x)$	Rule A
{1}	(2)	$H(y)$	ES, (1)
{3}	(3)	$(x)(H(x) \to M(x))$	Rule A
{3}	(4)	$H(y) \to M(y)$	US, (3)
{1, 3}	(5)	$M(y)$	T, (2), (4)
{1, 3}	(6)	$(\exists x)M(x)$	EG, (5)

   Note: In step 2 the variable y has been introduced by ES. Hence a conclusion such as $(x)M(x)$ cannot follow from step 5.

4. Prove that $(\exists x)(P(x) \wedge Q(x)) \Rightarrow (\exists x)P(x) \wedge (\exists x)Q(x)$.

   Solution

{1}	(1)	$(\exists x)(P(x) \wedge Q(x))$	Rule A
{1}	(2)	$P(y) \wedge Q(y)$	ES, (1)
{1}	(3)	$P(y)$	T, (2)

{1}	(4)	$Q(y)$	T, (2)
{1}	(5)	$(\exists x)P(x)$	EG, (3)
{1}	(6)	$(\exists x)Q(x)$	EG, (4)
{1}	(7)	$(\exists x)P(x) \wedge (\exists x)Q(x)$	T, (5), (6)

Note that the converse of the above is not true.

(1)	$(\exists x)P(x) \wedge (\exists x)Q(x)$	Rule A
(2)	$(\exists x)P(x)$	T, (1)
(3)	$(\exists x)Q(x)$	T, (1)
(4)	$P(y)$	ES, (2)
(5)	$Q(z)$	ES, (3)

5. Show that from $(\exists x)(F(x) \wedge S(x)) \rightarrow (y)(M(y) \rightarrow W(y))$ and $(\exists y)(M(y) \wedge \sim W(y))$, the conclusion $(z)(F(z) \rightarrow \sim S(z))$ follows.

Solution

{1}	(1)	$(\exists y)(M(y) \wedge \sim W(y))$	Rule A
{1}	(2)	$M(z) \wedge \sim W(z)$	ES, (1)
{1}	(3)	$\sim (M(z) \rightarrow W(z))$	T, (2)
{1}	(4)	$(\exists y) \sim (M(y) \rightarrow W(y))$	EG, (3)
{1}	(5)	$\sim (y)(M(y) \rightarrow W(y))$	(4)
{6}	(6)	$(\exists x)(F(x) \wedge S(x)) \rightarrow$ $(y)(M(y) \rightarrow W(y))$	Rule A
{1, 6}	(7)	$\sim (\exists x)(F(x) \wedge S(x))$	T, (5), (6)
{1, 6}	(8)	$(x) \sim (F(x) \wedge S(x))$	T, (7)
{1, 6}	(9)	$\sim (F(z) \wedge S(z))$	US, (8)
{1, 6}	(10)	$F(z) \rightarrow \sim S(z)$	T, (9)
{1, 6}	(11)	$(z)(F(z) \rightarrow \sim S(z))$	UG, (10)

6. Show that $(x)(P(x) \vee Q(x)) \Rightarrow (x)P(x) \vee (\exists x)Q(x)$.

Solution We will show this using indirect method of proof. In other words, we will assume $\sim ((x)P(x) \vee (\exists x)Q(x))$ as an additional premise.

{1}	(1)	$\sim ((x)P(x) \vee (\exists x)Q(x))$	Rule A (assumed)
{1}	(2)	$\sim (x)P(x) \wedge \sim ((\exists x)Q(x))$	T, (1)
{1}	(3)	$\sim (x)P(x)$	T, (2)
{1}	(4)	$(\exists x) \sim P(x)$	T, (3)

{1}	(5)	$\sim (\exists x) Q(x)$	T, (2)
{1}	(6)	$(x) \sim Q(x)$	T, (5)
{1}	(7)	$\sim P(y)$	ES, (4)
{1}	(8)	$\sim Q(y)$	US, (6)
{1}	(9)	$\sim P(y) \wedge \sim Q(y)$	T, (7), (8)
{1}	(10)	$\sim (P(y) \vee Q(y))$	T, (9)
{11}	(11)	$(x)(P(x) \vee Q(x))$	Rule A
{11}	(12)	$P(y) \vee Q(y)$	US, (11)
{1, 11}	(13)	$\sim (P(y) \vee Q(y)) \wedge (P(y) \vee Q(y))$	T, (10), (12)

7. Prove that $\sim P(a, b)$ follows logically from $(x)(y)(P(x, y) \rightarrow W(x, y))$ and $\sim W(a, b)$.

Solution

{1}	(1)	$(x)(y)(P(x, y) \rightarrow W(x, y))$	Rule A
{1}	(2)	$(y)(P(a, y) \rightarrow W(a, y))$	US, (1)
{1}	(3)	$P(a, b) \rightarrow W(a, b)$	US, (2)
{4}	(4)	$\sim W(a, b)$	Rule A
{1, 4}	(5)	$\sim P(a, b)$	T, (3), (4)

8. Prove that $\sim ((\exists x) P(x) \wedge Q(x)) \Rightarrow (\exists x)(P(x) \rightarrow \sim Q(x))$.

Solution

{1}	(1)	$\sim ((\exists x) P(x) \wedge Q(x))$	Rule A
{1}	(2)	$(x) \sim (P(x) \wedge Q(x))$	T, (1)
{1}	(3)	$\sim (P(y) \wedge Q(y))$	US, (2)
{1}	(4)	$P(y) \rightarrow \sim Q(y)$	T, (3)
{1}	(5)	$(\exists x)(P(x) \rightarrow \sim Q(x))$	EG, (4)

9. Prove that $(x)(P(x) \vee Q(x))$ and $(x) \sim P(x) \Rightarrow (x) Q(x)$.

Solution

{1}	(1)	$((x) P(x) \vee (\exists x) Q(x))$	Rule A
{1}	(2)	$P(y) \vee Q(y)$	US, (1)
{1}	(3)	$\sim P(y) \rightarrow Q(y)$	T, (2)
{4}	(4)	$(x) \sim P(x)$	Rule A
{4}	(5)	$\sim P(y)$	US, (4)
{1, 4}	(6)	$Q(y)$	T, (3), (5)
{1, 4}	(7)	$(x) Q(x)$	UG, (6)

10. Prove that $(x)(P(x) \rightarrow Q(x))$ and $(x)(R(x) \rightarrow \sim Q(x)) \Rightarrow (x)(R(x) \rightarrow \sim P(x))$.

Solution

{1}	(1)	$(x)(P(x) \rightarrow Q(x))$	Rule A
{1}	(2)	$P(y) \rightarrow Q(y)$	US, (1)
{3}	(3)	$(x)(R(x) \rightarrow \sim Q(x))$	Rule A
{3}	(4)	$R(y) \rightarrow \sim Q(y)$	US, (3)
{1}	(5)	$\sim Q(y) \rightarrow \sim P(y)$	T, (2)
{1, 3}	(6)	$R(y) \rightarrow \sim P(y)$	T, (4), (5)
{1, 3}	(7)	$(x)(R(x) \rightarrow \sim P(x))$	UG, (6)

11. Prove that $(\exists x)P(x) \rightarrow (x)Q(x)$ and $(x)(P(x) \rightarrow Q(x))$ are equivalent.

Solution

{1}	(1)	$(\exists x)P(x) \rightarrow (x)Q(x)$	Rule A
{1}	(2)	$\sim ((\exists x)P(x)) \vee (x)Q(x)$	T, (1)
{1}	(3)	$(x) \sim P(x) \vee (x)Q(x)$	T, (2)
{1}	(4)	$(x)(\sim P(x) \vee Q(x))$	T, (3)
{1}	(5)	$(x)(P(x) \rightarrow Q(x))$	T, (4)

Reversing the steps, we obtain the converse.

12. Prove that $\sim (x)(P(x) \wedge Q(x))$ and $(x)P(x) \Rightarrow \sim (x)Q(x)$.

Solution

{1}	(1)	$\sim (x)(P(x) \wedge Q(x))$	Rule A
{1}	(2)	$(\exists x) \sim (P(x) \wedge Q(x))$	T, (1)
{1}	(3)	$\sim (P(y) \wedge Q(y))$	ES
{1}	(4)	$\sim P(y) \vee \sim Q(y)$	T, (3)
{1}	(5)	$P(y) \rightarrow \sim Q(y)$	T, (4)
{6}	(6)	$(x)P(x)$	Rule A
{6}	(7)	$P(y)$	UG, (6)
{1}	(8)	$\sim Q(y)$	T, (5)
{1}	(9)	$(\exists x) \sim Q(x)$	EG, (8)
{1}	(10)	$\sim (x)Q(x)$	T, (9)

Exercises

1. Using the truth table or otherwise, check whether each of the following is a tautology.

(i) $(p \wedge q) \rightarrow p$
(ii) $(p \vee q) \rightarrow q$
(iii) $p \rightarrow (p \vee q)$
(iv) $q \rightarrow (p \vee q)$
(v) $\sim p \rightarrow (p \rightarrow q)$
(vi) $\sim (p \rightarrow q) \rightarrow p$
(vii) $(p \wedge (p \rightarrow q)) \rightarrow q$
(viii) $(\sim p \wedge (p \vee q)) \rightarrow q$
(ix) $(\sim q \wedge (p \rightarrow q)) \rightarrow \sim p$
(x) $((p \rightarrow q) \wedge (q \rightarrow r)) \rightarrow (p \rightarrow r)$

2. In the following problems, state whether the arguments given are valid or invalid. If they are valid, identify the tautology or tautologies used.

(i) If I drive to work, then I will arrive tired.
I will not be tired when I arrive at work.
∴ I do not drive to work.

(ii) If I drive to work, then I will arrive tired.
I arrive tired.
∴ I drive to work.

(iii) If I drive to work, then I will arrive tired.
I do not drive to work.
∴ I do not arrive tired.

Answer Invalid.

(iv) If I drive to work, then I will arrive tired.
I drive to work.
∴ I arrive tired.

(v) I will become famous or I will not become a writer.
I will become a writer.
∴ I will become famous.

Solution Let f stand for "I will become famous" and w stand for "I will become a writer". Then the first statement can be represented as $f \vee \sim w$. The second statement is w. Since both statements are assumed, the tautology is $(f \vee \sim w) \wedge w \rightarrow f$.

(vi) I will become famous or I will not become a writer.
I will not become a writer.
∴ I will become famous.

(vii) If I try hard and I have the talent, then I will become a musician.
If I become a musician, then I will be happy.
∴ If I will not be happy, then I did not try hard or I do not have the talent.

3. For each of the following, indicate the variables which are free and bound. Also, show the scope of the quantifiers.

   (i) $(x)(p(x) \wedge r(x)) \to (x)p(x) \wedge r(x)$
   (ii) $(x)(p(x) \wedge (\exists x)q(x)) \vee ((x)p(x) \to q(x))$

4. If the universe of discourse is the set $\{a, b, c\}$, eliminate the quantifiers in the following formulae.

   (i) $(x)p(x)$
   (ii) $(x)r(x) \wedge (x)s(x)$
   (iii) $(x)r(x) \wedge (\exists x)s(x)$
   (iv) $(x)(p(x) \to q(x))$
   (v) $(x) \sim p(x) \vee (x)p(x)$

5. Find the truth value of each of the following.

   (i) $(x)(p(x) \vee q(x))$, where $p(x) : x = 1, q(x) : x = 2$ and the universe of discourse is $\{1, 2\}$.
   (ii) $(x)(p \to q(x)) \vee r(a)$, where $p : 2 > 1, q(x) : x \leq 3, r(x) : x > 5, a : 5$ and the universe of discourse is $\{-2, 3, 6\}$.
   (iii) $(\exists x)(p(x) \to q(x)) \wedge T$, where $p(x) : x > 2, q(x) : x = 0, T$ is any tautology and the universe of discourse is $\{1\}$.

6. Show that $(\exists z)(q(z) \wedge r(z))$ is not implied by the formulas $(\exists x)(p(x) \wedge q(x))$ and $(\exists y)(p(y) \wedge r(y))$ by assuming a universe of discourse consisting of two elements.

7. Prove the following.

   (i) $\sim ((\exists x)p(x) \wedge q(a)) \Rightarrow (\exists x)p(x) \to \sim q(a)$
   (ii) $(x)(\sim p(x) \to q(x)), (x) \sim q(x) \Rightarrow p(a)$
   (iii) $(x)(p(x) \to q(x)), (x)(q(x) \to r(x)) \Rightarrow p(x) \to r(x)$
   (iv) $(x)(p(x) \vee q(x)), (x) \sim p(x) \Rightarrow (\exists x)q(x)$
   (v) $(x)(p(x) \vee q(x)), (x) \sim p(x) \Rightarrow (x)q(x)$
   (vi) $\sim (x)(p(x) \wedge q(x)), (x)p(x) \Rightarrow \sim (x)q(x)$
   (vii) $(\exists x)p(x) \to (x)q(x) \Rightarrow (x)(p(x) \to q(x))$
   (viii) $(x)(p(x) \to q(x)), (x)(r(x) \to \sim q(x)) \Rightarrow (x)(r(x) \to \sim p(x))$
   (ix) $(x)(p(x) \to q(x)) \Rightarrow (x)p(x) \to (x)q(x)$

8. Prove that the set $\{\sim, \to\}$ is functionally complete.

9. Define a connective $\leftarrow$ as follows.

p	q	$p \leftarrow q$
T	T	T
T	F	T
F	T	F
F	F	T

Prove that the set $\{\sim, \leftarrow\}$ is functionally complete.

2 Fundamentals of Set Theory

2.1 Introduction

Sets have been in use from time immemorial. Reference to sets crop up commonly in our use of language, for e.g., "A group (or a set) of people from Chennai came to our house yesterday". The concept of set theory, though very old, is very powerful and encompasses some profound results like the continuum hypothesis along with the elementary aspects that we are more familiar with.

In this chapter we will define a set and consider the various operations which can be performed on sets. We will prove a number of properties involving these operations.

2.2 Definition of a Set

We come across several instances where it so happens that we are very familiar with an object and can give elaborate description of the same, but asked to give a formal definition, we find it a puzzlingly difficult task. Set is a case in point. Intuitively we know many examples of sets but any attempt to give a universally accepted definition of a set will turn out to be an unsatisfactory task. We will content ourselves by saying that a set is a collection of well-defined objects. The objects in a set are called the elements of the set. Sets are normally denoted by capital letters while elements of a set are normally denoted by small letters. If x is an element of a set A, we write $x \in A$. If x is not an element of A then we write $x \notin A$.

Examples of sets

(a) Set S of all B.E. third semester students of Information Science and Engineering registered at BIET (Bapuji Institute of Engineering and Technology) during the academic year 2000–01.

(b) Set A of all classrooms in BIET. Each classroom in BIET is an element of this set.
(c) Set of all natural numbers (1, 2, 3, ...) denoted by N.
(d) Set of all integers (..., −3, −2, −1, 0, 1, 2, 3, ...) denoted by Z.
(e) Set of all rational numbers denoted by Q.
(f) Set of all irrational numbers denoted by I.
(g) Set of all real numbers denoted by $\mathbb{R}$.
(h) Set of all complex numbers denoted by C.

Note

1. Throughout this book, $N, Z, Q, I, \mathbb{R}$ and C will denote the sets with elements as mentioned above unless otherwise specified.
2. A rational number is a number of the form $\frac{a}{b}$ where a and b are integers and b is not equal to 0. Every integer is also a rational number (if n is an integer, then $n = \frac{n}{1}$). But there are rational numbers which are not integers. For example, $\frac{1}{2}$ is a rational number which is not an integer. Clearly, every rational number is a real number but there are several real numbers (for example $\sqrt{2}$) which are not rational numbers. A real number which is not a rational number is called an irrational number. A complex number is a number of the form $a + bi$ where a and b are real numbers and $i = \sqrt{-1}$ (imaginary number). If r is a real number, then we can write r as $r + 0i$ so that r is also a complex number. Thus every real number is also a complex number but there are many complex numbers (for example, $2 + 3i$) which are not real numbers.

2.3 Finite and Infinite Sets

Consider examples (a) and (b) given above. Naturally, we can say how many students of Information Science and Engineering registered in BIET during 2000–01 are in their third semester. We can also say how many classrooms there are in BIET. If we consider the set of all people who are in India during a given period, there is a definite number. Only, this number is much larger when compared to the two sets considered above. These are some examples of finite sets. A *finite set* is a set which contains only a finite number of elements. This number may be a small number like 5 or a large number like 20000 or a very large number like 800000584. The number of elements in a finite set A is denoted by $|A|$ and is called its *cardinality* or *order*.

Now consider the example (c) given above. We cannot say how many natural numbers there are or which is the largest natural number. If somebody says n is the largest natural number, we can add 1 to n and obtain a natural number $n + 1$

which is larger than n. We say the set of all natural numbers N is an infinite set. A set which is not finite is called an *infinite set*. Sets (c)–(h) given above are infinite sets.

2.4 Representation of Sets

A set is normally represented using curl brackets. If a set is finite and has only a small number of elements, then all the elements can be placed within curl brackets, separated by commas. For example, the set of all prime numbers between 2 and 15 can be represented by $\{2, 3, 5, 7, 11, 13\}$. This representation is not possible for infinite sets. Even if a set is finite but consists of many elements, it would be cumbersome (though not impossible) to represent the set by listing all its elements. For example, consider representing the set of all odd numbers less than 100000 which is certainly a finite set. Another method of representing a set is to take a variable and specify a property which the variable should satisfy so that it belongs to the set. This representation has the general form $\{x|x \text{ has } P\}$ where P is the name of a property. Using this representation, the set of all odd numbers less than 100000 can be represented as $\{x \in \ . \,|x \text{ is odd and } x < 100000\}$.

2.5 Subsets

Consider the set A of all classrooms in BIET. Let B denote the set of all rooms in BIET. B will include the principal's room, office room, staff rooms, conference room, etc., besides of course all the classrooms. Thus every element of A (classroom) is also an element of B (rooms in BIET). We say that A is a subset of B or A is contained in B and denote it by writing $A \subseteq B$. We also say that B is a superset of A or B contains A and denote it by writing $B \supseteq A$. We also note that there are rooms in B (for example staff rooms) which are not classrooms and hence do not belong to A. A is said to be a proper subset of B or A is said to be properly contained in B. This is denoted by writing $A \subset B$.

Definition A set X is said to be a *subset* of another set Y denoted by $X \subseteq Y$ if every element of X is also an element of Y. If X is a subset of Y and if there exists at least one element in Y which does not belong to X, then X is said to be a *proper subset* of Y.

Example (a) $N \subset Z$ (b) $Z \subset Q$ (c) $Q \subset \mathbb{R}$ (d) $\mathbb{R} \subset C$

Problem Prove that if $A \subset B$ and $B \subset C$, then $A \subset C$.

Note From the above problem, it follows that $N \subset Q, N \subset \mathbb{R}, N \subset C, Z \subset \mathbb{R}, Z \subset C$ and $Q \subset C$.

2.6 Equality of Sets

Two sets A and B are said to be *equal* if they contain exactly the same elements. If A and B are equal we write $A = B$.

Note The order in which elements appear in a set is not important. Thus if $A = \{$mango, orange, grape$\}$, $B = \{$orange, grape, mango$\}$ and $C = \{$grape, orange, mango$\}$, then $A = B = C$.

Problem For any two sets A and B, prove that $A = B$ if and only if $A \subseteq B$ and $B \subseteq A$.

Note In all such problems involving "if and only if", we have to first assume the left-hand side (what appears before the first if) and prove the right-hand side (what appears after the second if). Then assuming the right-hand side (what appears after the second if), we have to prove the left-hand side (what appears before the first if).

2.7 Universal Set

For any set A, we can always find a set U for which A is a subset. Then U is called a *universal set*. Consider the example (a) of Section 2.2. One universal set that we can consider for S is the set of all people in the world, which is obviously too large a set. Another universal set we can think of is the set of all the people in India, which is also a large set. A reasonable universal set for S can be the set of all students who have registered in BIET during 2000–2001, or a still better universal set can be the set of all third semester students who have registered in BIET during 2000–2001.

2.8 Venn Diagrams

Consider a set A and a universal set U. We know that $A \subset U$. The sets A and U can be represented using Venn diagrams as shown in Fig. 2.1.

In this diagram, a is an element of A and hence also an element of U. x is an element of U that is not in A.

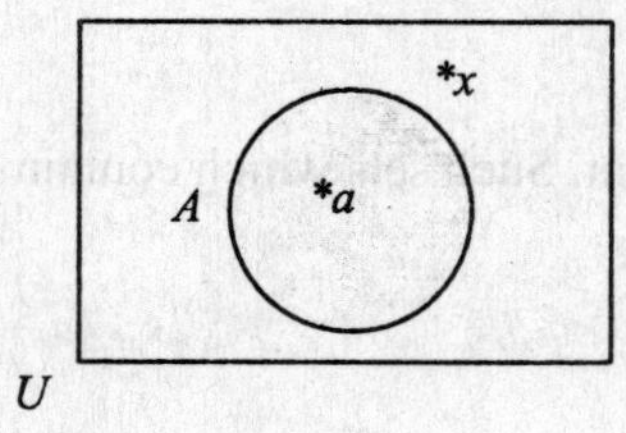

Fig. 2.1

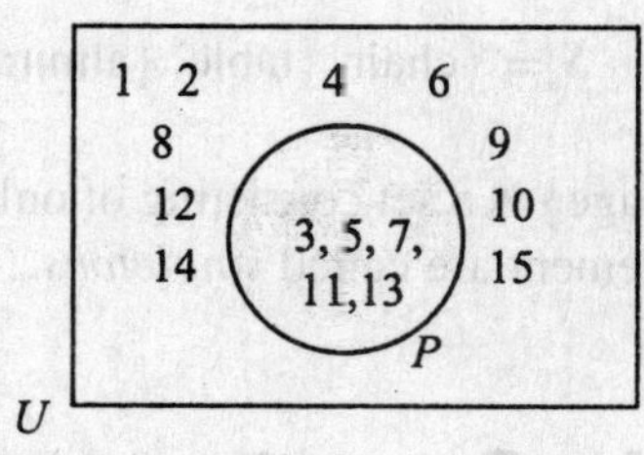

Fig. 2.2

Example Consider the set $P = \{p | p \text{ is prime and } 2 < p < 15\}$.

That is, $P = \{3, 5, 7, 11, 13\}$. One universal set U that we can consider for P is the set of all numbers from 1 to 15. $U = \{1, 2, 3, 4, \ldots, 15\}$.

The sets P and U can be represented using Venn diagram as given in Fig. 2.2.

2.9 Empty Set

Many a time we come across sets that do not contain any element. For example, consider the set of all prime numbers between 8 and 10. There is no such prime number and hence this set is empty. Another example of an empty set is the set of all people who are more than 300 years old. Again we know that there can be no such person and hence this set is also empty. Empty sets are denoted by $\{\}$ or ϕ. By convention, empty set is a subset of every set. Thus for any set X, ϕ and X are always subsets called trivial subsets. Any subset of X other than ϕ and X is called a non-trivial subset.

2.10 Power Set

We now consider sets whose elements themselves are sets. For a set X, consider the set whose elements are the various subsets of X. This set is called the power set of X and is denoted by $P(X)$. We know that $\phi \in P(X)$ and $X \in P(X)$.

Example Let $X = \{1, 2, 3\}$. Then $P(X) = \{\phi, \{1, 2, 3\}, \{1\}, \{2\}, \{3\}, \{1, 2\}, \{2, 3\}, \{1, 3\}\}$.

Problem For each of the sets S given below, find $P(S)$.

(i) $S = \{\}$
(ii) $S = \{\text{Peter}\}$
(iii) $S = \{\text{beans}, \{\text{cabbage}\}\}$

(iv) $S = \{\text{chair}, \{\text{table}, \{\text{almirah}\}\}, \text{stool}\}$

{cabbage} is a set consisting of only one element. Such sets which contain only one element are called *singletons*.

2.11 Operations on Sets

We know that if two numbers are given, we can perform various arithmetic operations on these numbers and obtain new numbers. For example, we can add the two numbers, multiply them and subtract one number from the other. When these operations are performed on two numbers, we get new numbers. These operations can also be extended to more than two numbers. Likewise, there are many operations which can be performed on two sets; these can also be extended to more than two sets. When operations are performed on sets, they yield new sets. We now consider some of these operations.

Union of two sets

If A and B are two sets, then their union, denoted by $A \cup B$, is defined as the set of all elements which belong to either A or B or both.

$$A \cup B = \{x \mid x \in A \text{ or } x \in B\}.$$

Thus all elements in A are in $A \cup B$, all elements in B are in $A \cup B$ and all elements in both A and B are also in $A \cup B$.

Exercises Prove the following.

(i) $A \subseteq A \cup B$
(ii) $B \subseteq A \cup B$

Example Let $X = \{\text{bread, rice, dal, potato}\}$ and $Y = \{\text{wheat, bread, cucumber, dal}\}$.

Then $X \cup Y = \{\text{bread, rice, dal, potato, wheat, cucumber}\}$.

Bread and dal appear both in X and Y. Hence they appear only once in $X \cup Y$.

Important note Whenever we want to prove that $x \in A \cup B$, we have to prove that either $x \in A$ or $x \in B$. Whenever we want to prove that $x \notin A \cup B$, we have to prove that $x \notin A$ and $x \notin B$.

Intersection of two sets

If A and B are two sets, then their intersection denoted by $A \cap B$ is defined as the set of all elements which belong to both A and B. For the sets X and Y given above, $X \cap Y = \{\text{bread, dal}\}$. Rice $\notin X \cap Y$ since rice $\notin Y$. Wheat $\notin X \cap Y$ since wheat $\notin X$.

Definition Two sets A and B are said to be *disjoint* if $A \cap B = \phi$.

Important note Whenever we want to prove that $x \in A \cap B$, we have to prove that $x \in A$ and $x \in B$. Whenever we want to prove that $x \notin A \cap B$, we have to prove that x does not belong to at least one of A or B.

Difference of two sets

If A and B are two sets, then $A - B$ is defined as the set of all elements which belong to A but not to B. $A - B$ is obtained by removing all those elements of A which also belong to B.

$$A - B = \{x \mid x \in A \text{ and } x \notin B\}.$$

Suppose A and B are disjoint sets. Then $A \cap B = \phi$. In this case, what is $A - B$? Obviously $A - B \subseteq A$. If $x \in A$, then $x \notin B$ so that $x \in A - B$ and hence $A \subseteq A - B$. We thus see that $A - B = A$.

Examples

1. Consider the sets X and Y given above. We have $X - Y = \{\text{rice, potato}\}$. Wheat $\notin X - Y$ since wheat $\notin X$. Though bread $\in X$, it also belongs to Y. Hence it does not belong to $X - Y$.
2. Let $A = \{\text{tiger, rice, parrot}\}$ and $B = \{\text{lion, apple}\}$. Since $(A \cap B) = \phi$, we have $A - B = A$.

Important note Whenever we want to prove that $x \in A - B$, we have to prove that $x \in A$ and $x \notin B$. Whenever we want to prove that $x \notin A - B$, we have to prove either $x \notin A$ or $x \in B$.

Complement of a set

Take $A = U$ in $A - B$. Then we have $U - B = \{x \mid x \in U \text{ and } x \notin B\} = \{x \notin B\}$. This set is called complement of B and is denoted by B^c. B^c denotes the set of all elements (which belong to the universal set) which do not belong to B.

Symmetric difference of two sets

If A and B are two sets, then the symmetric difference of A and B is denoted by $A \oplus B$ and is defined as

$$A \oplus B = (A - B) \cup (B - A).$$

It can be seen that $A \oplus B = (A \cup B) - (A \cap B)$.

2.12 Venn Diagram Representation of Operations on Sets

Consider two sets A and B. Let U be a universal set. Then the various operations on A and B are represented by the shaded portions in Fig. 2.3

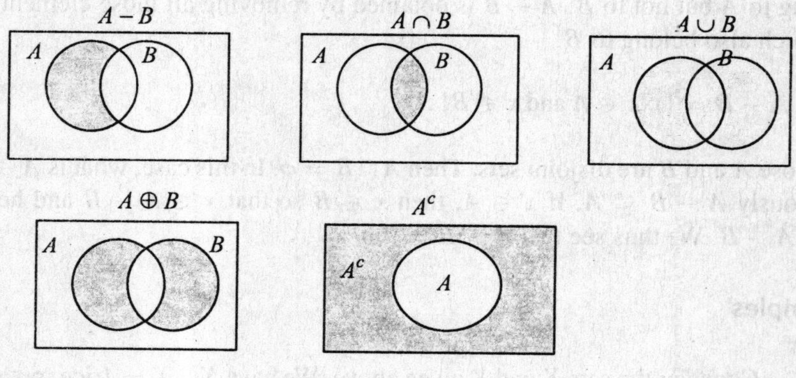

Fig. 2.3

2.13 Properties of Operations on Sets

(a) Union ($\cup$) operation is commutative. This means $A \cup B = B \cup A$ for any two sets A and B.

(b) Intersection ($\cap$) operation is commutative. This means $A \cap B = B \cap A$.

(c) Union ($\cup$) operation is associative. In other words, for any three sets A, B and C, $A \cup (B \cup C) = (A \cup B) \cup C$. Hence we need not use brackets and can simply write the left-hand side and the right-hand side as $A \cup B \cup C$.

(d) Intersection ($\cap$) operation is associative. In other words, for the three sets A, B and C, $A \cap ((B \cap C)) = (A \cap B) \cap C$. We can represent the left-hand side and the right-hand side as $A \cap B \cap C$ by omitting brackets.

(e) $A \cup \phi = A$ for any set A.
(f) $A \cap \phi = \phi$ for any set A.
(g) $A \cup U = U$.
(h) $A \cap U = A$.
(i) The operation of union is distributive with the operation of intersection. In other words, for any three sets A, B and C, we have

$$A \cup (B \cap C) = (A \cup B) \cap (A \cup C).$$

(j) The operation of intersection is distributive over the operation of union. In other words, for any three sets A, B and C, we have

$$A \cap (B \cup C) = (A \cap B) \cup (A \cap C).$$

(k) The union of a set and its complement is the universal set. In other words, for any set A, $A \cup A^c = U$. This means any element either belongs to A or belongs to A^c which is the same as saying that any element either belongs to A or does not belong to A.
(l) The intersection of a set and its complement is an empty set. In other words, for any set A, $A \cap A^c = \phi$. This means that there cannot be any element common to both A and its complement.
(m) Symmetric difference operation ($\oplus$) is commutative. This means $A \oplus B = B \oplus A$ for any two sets A and B.
(n) Symmetric difference operation ($\oplus$) is associative. In other words, for any three sets A, B and C, $A \oplus B \oplus C = (A \oplus B) \oplus C$.

From property (l) we note that A and A^c are disjoint sets. From property (c) we note that if we have n sets $A_1, A_2, \ldots, A_n$, then we can form their union $A_1 \cup A_2 \cup \cdots \cup A_n$. It is the set of all elements which belong to at least one of the sets $A_1, A_2, A_3, \ldots, A_n$. In other words, $A_1 \cup A_2 \cup \cdots \cup A_n = \{x | x \in A_i \text{ for some } i = 1, 2, 3, \ldots, n\}$.

From property (d) we note that if we have n sets $A_1, A_2, \ldots, A_n$, then we can form their intersection $A_1 \cap A_2 \cap \cdots \cap A_n$. It is the set of all elements which belong to all the sets $A_1, A_2, A_3, \ldots, A_n$. In other words, $A_1 \cap A_2 \cap \cdots \cap A_n = \{x | x \in A_i \text{ for all } i = 1, 2, 3, \ldots, n\}$.

2.14 De Morgan's Laws

For any two sets A and B, we have

1. $(A \cup B)^c = A^c \cap B^c$
2. $(A \cap B)^c = A^c \cup B^c$

Note that when we take complement, $\cup$ becomes $\cap$ and $\cap$ becomes $\cup$.

2.15 Cartesian Product

Before considering the definition of Cartesian product, we will look at a pair and an ordered pair. A pair consists of two components. For example, (a, b) is a pair where a and b can be any two objects. (mango, apple) is a pair which is the same as (apple, mango).

An ordered pair also has two components, but the difference between a pair and an ordered pair lies in the fact that in a pair (a, b) the order in which a and b appear does not matter whereas in an ordered pair (a, b), the ordering plays a very important role. As pairs, (a, b) is the same as (b, a), whereas as ordered pairs (a, b) is different from (b, a). For example, consider the points in a two-dimensional plane. We know that point (2, 8) is different from (8, 2) since the former has 2 as its x-coordinate and 8 as its y-coordinate, whereas the latter has 8 as its x-coordinate and 2 as its y-coordinate.

Definition Let A and B be two sets. The *Cartesian product* of A and B denoted by $A \times B$ is defined to be the set of all ordered pairs (a, b) where $a \in A$ and $b \in B$. In other words, $A \times B = \{(a, b) | a \in A, b \in B\}$.

Example Let A = {brinjal, leaf, mirchi} and B = {milk, coffee}. Then,

$$A \times B = \{\text{(brinjal, milk), (brinjal, coffee), (leaf, milk), (leaf, coffee),}$$
$$\text{(mirchi, milk), (mirchi, coffee)}\}.$$

$$B \times A = \{\text{(milk, brinjal), (milk, leaf), (milk, mirchi), (coffee, brinjal),}$$
$$\text{(coffee, leaf), (coffee, mirchi)}\}.$$

Since (brinjal, milk) is not the same as (milk, brinjal), it follows that $A \times B \neq B \times A$.

Problem If sets A and B have m and n elements respectively, what is the number of elements in (i) $A \times B$, (ii) $B \times A$?

We have $\mathbb{R} \times \mathbb{R} = \{(x, y) | x, y \in \mathbb{R}\}$. It is the set of all points in a two-dimensional plane and is denoted by $\mathbb{R}^2$.

2.16 Partition

Imagine a father who wants to distribute his property to his children. He gives a portion of his property to his first child, another portion to his second child and so on. No child is left out, which means that every child gets a portion of his property. Also, the same portion is not given to more than one child to avoid conflicts. If we put together the share of each child, we get the entire property of the father. In other words, the father has partitioned his property among his children.

Definition Let A be a set and let $A_1, A_2, A_3, \ldots\ldots\ldots\ldots, A_n$ be nonempty subsets of A. $A_1, A_2, A_3, \ldots\ldots\ldots\ldots, A_n$ are said to *partition* A if the following conditions are satisfied.

1. $A_i \neq \phi \; \forall i = 1, 2, 3, \ldots\ldots, n$.
2. $A_1 \cup A_2 \cup A_3 \cup \ldots\ldots\ldots \cup A_n = A$.
3. $A_1, A_2, A_3, \ldots\ldots, A_n$ are pairwise disjoint. This means $A_i \cap A_j = \phi$ whenever $i \neq j$.

Nonempty sets $A_1, A_2, A_3, \ldots\ldots\ldots\ldots, A_n$ form a partition of A if every element of A belongs to exactly one A_i. In this case, $A_1, A_2, A_3, \ldots, A_n$ are called the blocks of A.

Examples

1. Let $A = \{4, 9, 13, 18, 21, 29, 32\}$, $A_1 = \{4, 18, 31\}$, $A_2 = \{9, 21\}$ and $A_3 = \{13, 29\}$. Then $\{A_1, A_2, A_3\}$ forms a partition of A.
2. Consider the set A given above. Let $A_1 = \{4, 13, 18, 32\}$, $A_2 = \{9, 31\}$ and $A_3 = \{13, 29\}$. Then $\{A_1, A_2, A_3\}$ is not a partition of A since $A_1 \cap A_3 = \{13\}$ which is not empty.
3. Consider the set A given above. Let $A_1 = \{4, 18\}$, $A_2 = \{9, 21\}$ and $A_3 = \{13, 29\}$. Then $A_1 \cup A_2 \cup A_3 \neq A$ and hence $\{A_1, A_2, A_3\}$ is not a partition of A.

Now consider two finite sets A and B containing m and n elements respectively. $|A| = m$ and $|B| = n$. Suppose we want to find the order of $A \cup B$. This is not equal to $|A| + |B|$ since we have added the order of $A \cap B$ two times: once in the order of A and another in the order of B. Hence we should subtract one order of $A \cap B$ from $|A| + |B|$ to get the order of $A \cup B$. This means

$$|A \cup B| = |A| + |B| - |A \cap B|.$$

If A and B are disjoint sets, then order of $A \cap B$ is zero. That is, $|A \cap B| = 0$. In that case, $|A \cup B| = |A| + |B|$.

Example Let $A = \{\text{sugar, wheat, rice, dal}\}$ and $B = \{\text{dal, sugar, Horlicks, oil, jaggery}\}$.

Then $|A| = 4$ and $|B| = 5$.

$A \cup B = \{\text{sugar, wheat, rice, dal, Horlicks, oil, jaggery}\}$

$A \cap B = \{\text{sugar, dal}\}$

Hence $|A \cap B| = 2$ and $|A \cup B| = 7 = |A| + |B| - |A \cap B|$.

Solved problems

1. For any three sets A, B and C, prove that

$$|A \cup B \cup C| = |A| + |B| + |C| - |A \cap B| - |B \cap C| - |C \cap A| + |A \cap B \cap C|.$$

Solution Let $D = A \cup B$. Then,

$$\begin{aligned} |A \cup B \cup C| &= |D \cup C| = |D| + |C| - |D \cap C| \\ &= |A \cup B| + |C| - |(A \cup B) \cap C| \\ &= |A| + |B| - |A \cap B| + |C| - |(A \cap C) \cup (B \cap C)| \\ &= |A| + |B| - |A \cap B| + |C| - (|A \cap C| + |B \cap C| \\ &\quad - |(A \cap C) \cap (B \cap C)|) \\ &= |A| + |B| + |C| - |A \cap B| - |B \cap C| - |C \cap A| \\ &\quad + |A \cap B \cap C|. \end{aligned}$$

2. A survey of 500 people for their preference of TV channels produced the following information:
285 watch Sun TV, 195 watch Udaya, 115 watch Gemini, 45 watch both Sun TV and Gemini, 70 watch both Sun TV and Udaya, 50 watch both Udaya and Gemini. 50 do not watch any of the three channels. Find

   (a) How many people in the survey watch all the three channels ?
   (b) How many people watch exactly one of the three channels ?

Solution

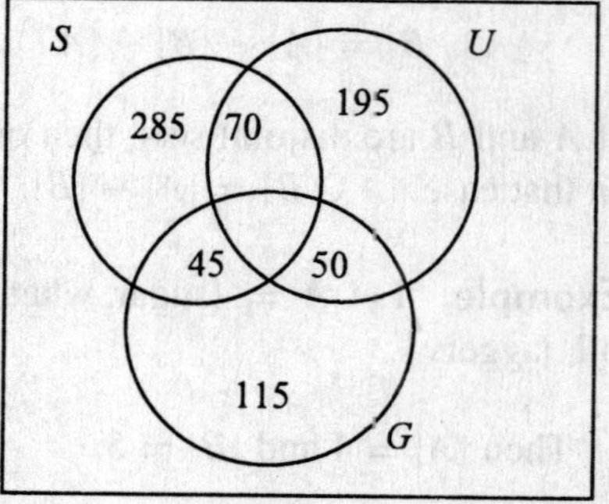

Let S, U and G denote the set of all people who watch Sun TV, Udaya and Gemini respectively. Let S_1, U_1 and G_1 denote the set of all people who watch only Sun TV, only Udaya and only Gemini respectively. It is given that $|S| = 285$, $|G| = 115$, $|U| = 195$, $|S \cap G| = 45$, $|S \cap U| = 70$, $|U \cap G| = 50$. Since it is given that of the total of 500 people who were surveyed, 50 do not watch any of the three channels, it follows that 450 people watch at least one of the three channels (note that these 450 people may watch two or three channels also). Hence $|S \cup U \cup G| = 450$.

(a) We know

$$|S \cup U \cup G| = |S| + |U| + |G| - |S \cap U| - |U \cap G| - |S \cap G| + |S \cap U \cap G|.$$

Substituting the various values, we obtain $450 = 285 + 195 + 115 - 70 - 50 - 45 + |S \cap U \cap G|$. Hence $|S \cap U \cap G| = 20$. This means 20 people watch all the three channels.

(b) We have

$$|S_1| = |S| - |S \cap U| - |S \cap G| + |S \cap U \cap G|$$

$$= 285 - 70 - 45 + 20 = 190.$$

$$|G_1| = |G| - |U \cap G| - |S \cap G| + |S \cap U \cap G|$$

$$= 115 - 50 - 45 + 20 = 40.$$

$$|U_1| = |U| - |U \cap G| - |S \cap U| + |S \cap U \cap G|$$

$$= 195 - 50 - 70 + 20 = 95.$$

Hence the number of people who watch exactly one of the three channels $= |S_1| + |G_1| + |U_1| = 190 + 40 + 95 = 325$.

3. For any three sets A, B and C, prove the following.
   (i) $A - (B - C) = (A - B) - C$ if and only if $A \cap C = \phi$
   (ii) $(A-B)-C = A-(B \cup C) = (A-C)-(B-C) = (A-C)-B$
   (iii) $(A \cap B) \cup C = A \cap (B \cup C)$ if and only if $C \subseteq A$.

Solution

(i) We will first assume that $A - (B - C) = (A - B) - C$ and we will prove that $A \cap C = \phi$. Let, if possible, $A \cap C \neq \phi$. Then there exists an element $x \in A \cap C$. Now $x \in A$ and $x \in C$. Hence by definition of $B - C$, $x \notin B - C$. By definition of $A - (B - C)$, $x \in A - (B - C) = (A - B) - C$ (by assumption). Again, by definition of $(A - B) - C$ it follows that $x \notin C$, a contradiction. Hence our assumption that $A \cap C \neq \phi$ is false. This proves that $A \cap C = \phi$.

Conversely, assuming that $A \cap C = \phi$, we will prove that $A - (B - C) = (A - B) - C$. Let $x \in A - (B - C)$. Then $x \in A$ and $x \notin B - C$. By definition of $B - C$, $x \notin B$ or $x \in C$. (Note this point !) If $x \in C$, then $x \in A \cap C = \phi$ (by assumption). Hence $x \notin C$. Now $x \notin B$ or $x \in C$ implies that $x \notin B$. Now $x \in A$ and $x \notin B$ means $x \in A - B$. Since $x \notin C$, it follows that $x \in (A - B) - C$

proving that $A - (B - C) \subseteq (A - B) - C$. Similarly, we can prove that $(A - B) - C \subseteq A - (B - C)$.

(ii) Let $x \in (A - B) - C$. Then $x \in A - B$ and $x \notin C$. $x \in A - B$ means $x \in A$ and $x \notin B$. Since $x \notin B$ and $x \notin C$, we have $x \notin B \cup C$. (Note this point !) Since $x \in A$ and $x \notin B \cup C$, it follows that $x \in A - (B \cup C)$ proving that $(A - B) - C \subseteq A - (B \cup C)$. Similarly, we can prove that $A - (B \cup C) \subseteq (A - B) - C$.

We will now prove that $A - (B \cup C) = (A - C) - (B - C)$. Let $x \in (A - C) - (B - C)$. Then $x \in A - C$ and $x \notin B - C$. $x \in A - C$ means that $x \in A$ and $x \notin C$. $x \notin B - C$ means that $x \notin B$ or $x \in C$. Since we already know $x \notin C$, we conclude $x \notin B$. Since $x \notin B$ and $x \notin C$, we have $x \notin B \cup C$. Since $x \in A$, it follows that $x \in A - (B \cup C)$ proving that $(A - C) - (B - C) \subseteq A - (B \cup C)$. Similarly, we can prove that $A - (B \cup C) \subseteq (A - C) - (B - C)$.

We will now prove that $(A - C) - (B - C) = (A - C) - B$. If $x \in (A - C) - (B - C)$, then $x \in A - C$ and $x \notin B - C$. $x \in A - C$ means $x \in A$ and $x \notin C$. $x \notin B - C$ means $x \notin B$ or $x \in C$. Since we already know $x \notin C$, we conclude $x \notin B$. Since $x \in A - C$ and $x \notin B$, it follows that $x \in (A - C) - B$ proving that $(A - C) - (B - C) \subseteq (A - C) - B$. Similarly, we can prove that $(A - C) - B \subseteq (A - C) - (B - C)$.

(iii) By first assuming that $(A \cap B) \cup C = A \cap (B \cup C)$, we will prove $C \subseteq A$. If $x \in C$, then $x \in (A \cap B) \cup C = A \cap (B \cup C)$, which means $x \in A$. Hence $C \subseteq A$.

Conversely, assuming that $C \subseteq A$, we will prove that $(A \cap B) \cup C = A \cap (B \cup C)$. Suppose $x \in (A \cap B) \cup C$. Then $x \in A \cap B$ or $x \in C$. If $x \in A \cap B$, then $x \in A$. Also $x \in B$ means $x \in B \cup C$ and hence $x \in A \cap (B \cup C)$. If $x \in C$, then $x \in A$ (since $C \subseteq A$). Also $x \in C$ means $x \in B \cup C$ and hence $x \in A \cap (B \cup C)$. Thus in both cases, we have proved that $(A \cap B) \cup C \subseteq A \cap (B \cup C)$.

If $y \in A \cap (B \cup C)$, then $y \in A$ and $y \in B \cup C$. Hence $y \in B$ or $y \in C$. If $y \in B$, then $y \in A \cap B$ so that $y \in (A \cap B) \cup C$. If $y \in C$, then clearly $y \in (A \cap B) \cup C$. In both cases, $A \cap (B \cup C) \subseteq (A \cap B) \cup C$.

Exercises

1. For any three sets A, B and C, given that $A \cap B = A \cap C$ and $A^c \cap B = A^c \cap C$, is it necessary that $B = C$? Justify your answer.
2. For any three sets A, B and C, prove the following.

(a) $A - B = A - (A \cap B)$

(b) If $A \cap C \subseteq B \cap C$ and $A \cap C^c \subseteq B \cap C^c$ then $A \subseteq B$.

3. Prove that symmetric difference is associative. For any three sets A, B and C, $(A \oplus B) \oplus C = A \oplus (B \oplus C)$.
4. In a class, 30 students took Mathematics, 35 students took Computer Science, 100 students took Electronics, 15 students took both Mathematics and Computer Science, 15 students took both Mathematics and Electronics, 20 students took Computer Science and Electronics and 5 students took all the three subjects. How many students took at least one of the three subjects ?
5. In a survey of 260 students, the following data were obtained. 60 had taken Mathematics, 94 had taken Computer Science, 58 had taken Business Studies, 28 had taken both Mathematics and Business Studies, 26 had taken both Mathematics and Computer Science, 22 had taken Computer Science and Business Studies, 14 had taken all the three courses. Find

   (a) How many students had taken none of the three courses ?

   (b) Of the students surveyed, how many had taken

      (i) only Computer Science course ?

      (ii) only Mathematics course ?

      (iii) only Business Studies course ?

6. Among 100 students, 32 study Mathematics, 20 study Physics, 45 study Bioiogy, 15 study both Mathematics and Biology, 7 study both Mathematics and Physics, 10 study both Physics and Biology, 30 do not study any of the three subjects. Find

   (a) The number of students who study all the three subjects.

   (b) The number of students who study exactly one of the three subjects.

7. A survey was conducted among 1000 people. Of these, 595 are Democrats, 595 wear glasses and 550 like ice cream; 395 of them are Democrats who wear glasses, 350 of them are Democrats who like ice cream and 400 of them wear glasses and like ice cream, 250 of them are Democrats who wear glasses and like ice cream. How many of them who are not Democrats do not wear glasses and do not like ice cream ? How many of them are Democrats who do not wear glasses and do not like ice cream ?
8. It is known that at a university, 60% of the professors play tennis, 50% of them play bridge, 70% jog, 20% play tennis and bridge, 30% play tennis and jog and 40% play bridge and jog. If someone claimed that

20% of the professors jog and play bridge and tennis, would you believe this claim ? Why ?

9. 60,000 fans who attended the finals of a one-day international match purchased all the paraphernalia on sale for promoting the event. Altogether, 20,000 T-shirts, 36,000 bats with monogram and 12,000 key rings were sold. We know that 52,000 fans bought at least one item and no one bought more than one of a given item. Also, 6,000 fans bought both bats and key rings, 9,000 bought both bats and T-shirts and 5,000 bought both key rings and T-shirts.

   (a) How many fans bought all three items?
   (b) How many fans bought exactly one item?
   (c) Someone questioned the accuracy of the total number of purchasers: 52,000 (given that all the other numbers have been confirmed to be correct). This person claimed the total number of purchasers to be either 60,000 or 44,000. How do you dispel the claim?

10. In a class of 130 students, 60 wore hats, 51 wore scarves and 30 wore both hats and scarves. Of the 54 students who wore sweaters, 26 wore hats, 21 wore scarves and 12 wore both hats and scarves. All those who did not wear either a hat or a scarf wore gloves.

   (a) How many students wore gloves ?
   (b) How many students not wearing a sweater wore hats but not scarves ?
   (c) How many students not wearing a sweater wore neither a scarf nor a hat ?

3 Principles of Mathematical Induction

Suppose we want to prove that a result is true for all values of n greater than or equal to n_0. In such cases we first prove that the result is true for n_0. This step is called the basis step. Then we assume that the result is true for $k(\geq n_0)$ and prove that the result is true for $k+1$. This step is called induction step. This will prove that the result is true for all values greater than or equal to n_0.

Note

1. If we want to prove that a result is true for values $\geq n_0$, we need not bother to see what happens to the result for values less than n_0.
2. The principle of mathematical induction can be applied only when the result is known.

Solved problems

1. Prove that if a set S has n elements, then $P(S)$ has 2^n elements.

   Solution
   Basis step: Suppose $n = 1$. Then S contains only one element, say a. Then the subsets of S are ϕ and $\{a\}$. Hence $P(S)$ contains $2 = 2^1$ elements. This proves the result for $n = 1$. If $n = 2$, then A contains two elements, say a and b. Then $P(S) = \{\phi, \{a\}, \{b\}, \{a, b\}\}$ so that $P(S)$ contains $4 = 2^2$ elements. This proves the result for $n = 2$.

   Induction step: Assume that whenever a set contains k elements, its power set contains 2^k elements and let $S = \{a_1, a_2, a_3, \ldots, a_k, a_{k+1}\}$ be a set containing $k+1$ elements. Let $B = \{a_1, a_2, a_3, \ldots, a_k\}$. By induction hypothesis B contains 2^k subsets. Let them be $B_1, B_2, B_3, \ldots, B_2{}^k$. Note that $B_1, B_2, B_3, \ldots, B_2{}^k$ are also subsets of A. This gives 2^k subsets of A. Now consider $B_1 \cup \{a_{k+1}\}, B_2 \cup \{a_{k+1}\}, \ldots, B_2{}^k \cup \{a_{k+1}\}$. No two of these sets are equal. For, if $B_i \cup \{a_{k+1}\} = B_j \cup \{a_{k+1}\}$, then $B_i = B_j$. Also, none of these sets is equal to any of $B_1, B_2, B_3, \ldots, B_2^k$.

Thus $B_1 \cup \{a_{k+1}\}, B_2 \cup \{a_{k+1}\}, \ldots, B_{2^k} \cup \{a_{k+1}\}$ give 2^k subsets of A. These 2^k subsets together with $B_1, B_2, B_3, \ldots, B_{2^k}$ give $2^k + 2^k = 2(2^k) = 2^{(k+1)}$ subsets of A. This completes the proof.

2. Prove the following using induction.

$$1 \times 1! + 2 \times 2! + 3 \times 3! + \cdots + n \times n! = (n+1)! - 1.$$

Solution
Basis step: If $n = 1$ then, the L.H.S of the equation $= 1 \times 1! = 1$ while the R.H.S $= (1+1)! - 1 = 1$, proving that the result is true for $n = 1$. If $n = 2$, then the L.H.S $= 1 \times 1! + 2 \times 2! = 5$ while the R.H.S $= (2+1)! - 1 = 5$, proving that the result is true for $n = 2$.

Induction step: Now assume that the result is true for $n = k$ which means

$$1 \times 1! + 2 \times 2! + 3 \times 3! + \cdots + k \times k! = (k+1)! - 1 \ldots \quad (3.1)$$

We will prove that

$$1 \times 1! + 2 \times 2! + 3 \times 3! + \cdots + k \times k! + (k+1) \times (k+1)! = ((k+1)+1)! - 1.$$

Now L.H.S $= 1 \times 1! + 2 \times 2! + 3 \times 3! + \cdots + k \times k! + (k+1) \times (k+1)!$

$$= (k+1)! - 1 + (k+1) \times (k+1)! \quad \text{(from (3.1))}$$

$$= (k+1)!(1 + k + 1) - 1$$

$$= (k+1)!(k+2) - 1.$$

$$= (k+2)! - 1$$

which is the R.H.S.

3. Prove that $2^n > n^3$ for $n \geq 10$.

Note The result may or may not be true when n is less than 10. When $n = 3$, we get $2^3 = 8$ which is not greater than $3^3 = 27$.

Solution
Basis step: If $n = 10$, then the L.H.S $= 2^{10} = 1024$ while the R.H.S $= 10^3 = 1000$. Hence the result is true when $n = 10$.

Induction step: Now assume that the result is true for $n = k (\geq 10)$. Then,

$$2^k > k^3 \text{ for } k \geq 10 \ldots \quad (3.2)$$

We have to prove that $2^{(k+1)} > (k+1)^3$.

$$
\begin{aligned}
\text{We have}\quad 2^{k+1} &= 2 \times 2^k \\
&> \left(1 + \frac{1}{10}\right)^3 \times 2^k \text{ since } 2 > \left(1 + \frac{1}{10}\right)^3 \\
&\geq \left(1+\frac{1}{k}\right)^3 \times 2^k \text{ since } k \geq 10 \text{ implies } \frac{1}{k} \leq \frac{1}{10}. \\
&> \left(1 + \frac{1}{k}\right)^3 \times k^3 \text{ (from (3.2))}. \\
&= (k+1)^3.
\end{aligned}
$$

4. Prove that any positive integer greater than or equal to 24 can be expressed as a sum of 5s and / or 7s.

Solution
Basis step: We have

$$
\begin{aligned}
&24 = 5 + 5 + 7 + 7, 25 = 5 + 5 + 5 + 5 + 5, \\
&26 = 5 + 7 + 7 + 7, 27 = 5 + 5 + 5 + 5 + 7 \\
\text{and}\quad &28 = 7 + 7 + 7 + 7.
\end{aligned}
$$

Thus the numbers 24, 25, 26, 27 and 28 can be expressed as a sum of 5s and / or 7s.

Induction step: Now assume that the result is true for any positive integer greater than or equal to 24 and less than k where $k \geq 28$. Let $n = k + 1$. We have to prove that the result is true for n. We have $n = k + 1 = (k - 4) + 5 = l + 5$ where $l = k - 4$. Obviously, $l < k$ and $l \geq 24$. By inductive hypothesis, l can be expressed as a sum of 5s and / or 7s and hence $n = l + 5$ can also be expressed as such a sum

Exercises

Prove the following using the principle of mathematical induction.

1. For $n \geq 1$, $1 + 2 + 3 + \cdots + n = \dfrac{n(n+1)}{2}$.
2. $n \geq 2^{(n-1)}$ for $n \geq 1$.
3. Show that any positive integer greater than or equal to 2 is either a prime number or a product of prime numbers.
4. $1^2 + 3^2 + 5^2 + \cdots + (2n-1)^2 = \dfrac{n(2n+1)(2n-1)}{3}$.

5. $1^3 + 2^3 + 3^3 + \cdots + n^3 = \dfrac{n^2(n+1)^2}{4}$.
6. $1 + 5 + 9 + \cdots + (4n - 3) = n(2n - 1)$.
7. $1 + 2^n < 3^n$, for $n \geq 2$.
8. $n < 2^n$ for $n \geq 1$.
9. $\dfrac{1}{(1 \times 2)} + \dfrac{1}{(2 \times 3)} + \dfrac{1}{(3 \times 4)} + \cdots + \dfrac{1}{n(n+1)} = \dfrac{n}{(n+1)}$.
10. $\dfrac{1}{(1 \times 3)} + \dfrac{1}{(3 \times 5)} + \dfrac{1}{(5 \times 7)} + \cdots + \dfrac{1}{((2n-1) \times (2n+1))} = \dfrac{n}{(2n+1)}$.
11. $\dfrac{1}{(1 \times 4)} + \dfrac{1}{(4 \times 7)} + \dfrac{1}{(7 \times 10)} + \cdots + \dfrac{1}{((3n-2) \times (3n+1))} = \dfrac{n}{(3n+1)}$.
12. Prove that any positive integer greater than or equal to 14 can be expressed as a sum of 3s and / or 8s.
13. $n^3 + 2n$ is divisible by 3 for all $n \geq 1$.
14. $n^4 - 4n^2$ is divisible by 3 for all $n \geq 2$.
15. $2^n \times 2^{n-1}$ is divisible by 3 for all $n \geq 1$.
16. $1 + 2 + 2^2 + 2^3 + \cdots + 2^n = 2^{n+1} - 1$.
17. For $n \geq 1$, $1^2 - 2^2 + 3^2 - 4^2 + \cdots + (-1)^{n-1}n^2 = (-1)^{n-1}\dfrac{n(n+1)}{2}$.
18. $\dfrac{1^2}{(1 \times 3)} + \dfrac{2^2}{(3 \times 5)} + \cdots + \dfrac{n^2}{(2n-1)(2n+1)} = \dfrac{n(n+1)}{(4n+2)}$.
19. $1 \times 2 \times 3 + 2 \times 3 \times 4 + 3 \times 4 \times 5 + \cdots + n(n+1)(n+2) = \dfrac{n(n+1)(n+2)(n+3)}{4}$.

4 Permutations and Combinations

Consider the following problem.

Suppose there are n letters using which we want to form words of length r. In how many ways can this be done if

1. Letters in the word are repeated ?
2. Repetitions are not allowed ?

We will consider question (1) first. If repetitions of letters are allowed, then we have n choices to choose the first letter in the word of length r, again n choices to choose the second letter, n choices to choose the third letter and finally n choices to choose the rth letter. Hence the total number of words that can be formed is $n \times n \times \cdots \times n(r \text{ times})$ which is n^r.

We will now consider question (2). If repetitions are not allowed, then there are n choices to choose the first letter in the word of length r, $n - 1$ choices to choose the second letter (note that the letter which was chosen as the first letter cannot be chosen again), $n - 2$ choices to choose the third letter and finally $n - (r - 1)$ choices to choose the rth letter. Hence the total number of words that can be formed is $n \times (n - 1) \times (n - 2) \cdots \times (n - r + 1)$.

Suppose tasks T_1 and T_2 have to be performed in sequence. If n_1 denotes the number of ways in which task T_1 can be performed and n_2 the number of ways in which task T_2 can be performed, then the sequence T_1 T_2 (task T_1 followed by task T_2) can be performed in $n_1 \times n_2$ ways. This can be argued as follows.

Each of the n_1 ways of performing task T_1 leads to different ways of performing the sequence T_1 T_2. To each of these n_1 ways, there exist n_2 ways of performing task T_2. Hence the number of ways in which the sequence T_1 T_2 can be performed is n_1 n_2. Generalizing this, we obtain the following.

If tasks $T_1, T_2, T_3, \ldots, T_k$ can be performed in $n_1, n_2, n_3, \ldots, n_k$ ways respectively, then the sequence $T_1 T_2 T_3 \ldots T_k$ can be performed in $n_1 n_2 n_3 \ldots n_k$ ways.

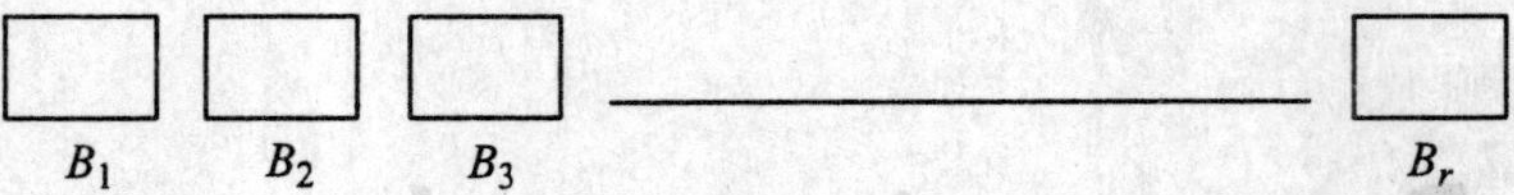

The formation of words of length r can be thought of as filling up the above boxes assuming that each box can hold only one letter.

Now consider again question (1).

There are n ways of filling box B_1 (any of the n letters can be placed in that box). Since repetition of letters is allowed, there are n ways of filling box B_2. Similarly there are n ways of filling box B_3 and so on. If filling the box can be thought of as performing a task, then the total number of ways in which the r boxes can be filled is $n \times n \times \cdots \times n$ which is n^r. Each way of filling up the r boxes gives rise to one word of length r and hence the number of words of length r that can be formed out of n letters is also equal to n^r.

Now consider question (2). Again there are n ways of filling box B_1. Since the letters in the word cannot be repeated, the letter that went into box B_1 cannot be placed in box B_2. Any of the remaining $(n-1)$ letters can be placed in box B_2. Any of the remaining $(n-2)$ letters (other than the two letters which were placed in box B_1 and box B_2) can be placed in box B_3. Repeating in this manner, the number of ways of filling box B_r is $(n-r+1)$. Hence the total number of ways in which the boxes can be filled is $n(n-1)(n-2)\cdots(n-r+1)$ which is also the number of words of length r, each with distinct letters that can be formed out of n letters.

Example Consider the five letters a, b, c, d, e. Determine the number of words of length 4 that can be formed out of these five letters if (i) repetition of letters is allowed, (ii) repetition of letters is not allowed. The words may or may not have a meaning.

Answer

(i) $5^4 = 625$

(ii) $5 \times 4 \times 3 \times 2 = 120$

If we consider a word of length r (with distinct letters chosen from n letters), it corresponds to an arrangement of r objects out of a total of n objects. Each letter can be thought of as an object. The number of ways in which r objects out of n objects can be arranged is called the number of permutations of n objects taken r at a time; this number is denoted by n_r^p. From the above discussion we note that $n_r^p = n(n-1)(n-2)\cdots(n-r+1)$; this can be written as

$$\frac{n(n-1)(n-2)\cdots(n-r+1)(n-r)(n-r-1)\cdots 3\times 2\times 1}{(n-r)(n-r-1)\cdots 3\times 2\times 1},$$

which is the same as $\frac{n!}{(n-r)!}$.

Suppose $r = n$. In this case, we are interested in the number of ways of forming words of length n (each word with distinct letters) out of n letters. This is nothing but the number of ways in which the n letters can be arranged, which is $n(n-1)(n-2)\cdots 3\times 2\times 1 = n!$. Each way of arranging the n letters is a permutation on n letters. It follows that the number of permutations on n letters is $n!$

For example, if $A = \{a, b, c\}$, then the various permutations on A are $abc, acb, bac, bca, cab, cba$. The number of permutations on A is $6 = 3!$ If $B = \{a, b, c, d\}$, then there will be $4! = 24$ permutations on B. Each of these 24 permutations will give rise to a word of length 4.

Combinations

Consider a set A consisting of n elements. Any subset of A containing r elements is called an r combination or a combination of n symbols taken r at a time. The number of subsets of A each containing r elements is called the number of combinations of n elements taken r at a time.

Assume there are 5 students S_1, S_2, S_3, S_4 and S_5 and we want to form a committee consisting of 3 students. The various possibilities are as follows. $\{S_1, S_2, S_3\}, \{S_1, S_4, S_5\}, \{S_1, S_2, S_4\}, \{S_2, S_3, S_4\}, \{S_1, S_2, S_5\}, \{S_2, S_3, S_5\}, \{S_1, S_3, S_4\}, \{S_2, S_4, S_5\}, \{S_1, S_3, S_5\}$ and $\{S_3, S_4, S_5\}$.

There are 10 ways in which the committee can be formed.

Note

1. Since we have included $\{S_1, S_3, S_5\}$, we do not include $\{S_3, S_5, S_1\}$, $\{S_1, S_5, S_3\}$, etc. After all, $\{S_1, S_3, S_5\}$ says we have included the students S_1, S_3 and S_5 while $\{S_3, S_5, S_1\}$ says we have included the students S_3, S_5 and S_1. Both are one and the same.
2. In a permutation the order of elements is of consequence whereas in a combination the order does not matter.

Consider the problem of selecting r objects out of n objects. The r objects we select can be thought of as a subset consisting of r elements. It follows that the number of ways in which r objects can be selected out of n objects is also the number of combinations of n objects taken r at a time. It is denoted by n_{C_r}.

Consider a subset consisting of r elements from a set A containing n elements. This one subset gives rise to $r!$ permutations of n symbols taken r at a time.

(Note that there are r ways of filling up the first position, $r-1$ ways of filling up the second position and so on). We know that the number of subsets of A (each subset containing r elements) is ${}^nC_r$. Hence the total number of permutations of n symbols taken r at a time that can arise from ${}^nC_r$ subsets is $r! \times {}^nC_r$. We also know that the number of permutations of n symbols taken r at a time is ${}^nP_r$. Hence we obtain $r! \times {}^nC_r = {}^nP_r$ so that

$$nC_r = \frac{{}^nP_r}{r!} = \frac{n!}{((n-r)! \times r!)}.$$

Exercises

1. A bank password consists of two letters of English alphabet followed by two digits from 0 to 9. How many different passwords are there ?
2. A coin is tossed 4 times and the result of each toss is recorded. How many different sequences of heads and tails are possible ?
3. A die is tossed 4 times and the numbers shown are arranged in a sequence. Determine the number of different sequences.
4. In how many ways can 6 married couples be seated in a row if .
   (a) any person can sit next to any other person ?
   (b) men and women occupy alternate seats ?
5. Find the number of different permutations of the letters in the word GROUP.
6. In how many ways can a committee of 3 faculty members and two students be selected from 7 faculty members and 8 students ?
7. An organization has decided to appoint for each floor one male and one female residential advisor. How many different pairs of advisors can be selected for a building with 7 floors from 12 male and 15 female candidates ?
8. A box contains 15 balls 8 of which are red and 7 are black. In how many ways can 5 balls be chosen so that
   (i) All 5 are black ?
   (ii) Two are red and three are black?
   (iii) Three are red and two are black ?
9. A student must answer 8 out of 10 questions in an examination.
   (a) How many choices does the student have ?
   (b) How many choices does he / she have if the first three questions are compulsory ?
   (c) How many choices does the student have if four out of the first five questions are compulsory ?

5 Recurrence Relations

5.1 Introduction

Suppose you ask a man his age. There are several ways in which the man can tell you his age. He may tell you directly that his age is 25, or instead he may say that he is four years older than his sister. Your immediate question next will be "What is the age of your sister ?" If he tells the age of his sister directly, you can add 4 to it and get the age of the man. Instead of telling the age of his sister directly, he may say that she is nine years younger than his brother. Your next question will be "What is the age of your brother?" This process can go on. Ultimately, the man has to give one age using which you can determine his age. This is exactly how a recurrence relation works.

Many times we define the elements of a sequence in terms of its previous elements which will enable us to express the elements of a sequence in a very compact form. Consider the sequence $(1, 5, 9, 13, \ldots)$. We note that every term of the sequence from the 2nd term onwards is 4 more than the previous term. If we know the first term, we can determine the other terms of the sequence. It is not possible to determine other terms of the sequence if the first term is not known. A recurrence relation for this sequence is

$$a_n = a_{n-1} + 4(n > 1).$$

The above equation has no meaning when $n = 1$.

Another familiar example is the Fibonacci sequence in which every element from the third element onwards is the sum of the previous two terms. If f_n, f_{n-1}, f_{n-2} denote the nth term, $(n-1)$th term and $(n-2)$th term respectively, then the recurrence relation becomes

$$f_n = f_{n-1} + f_{n-2}.$$

Naturally, $n > 2$. (If $n \leq 2$, then f_{n-2} will not have any meaning). Obviously, this definition of Fibonacci sequence will not be complete if we do not specify the first two elements f_1 and f_2. If f_1 and f_2 are not known, then f_3 which is the sum of f_1 and f_2 cannot be determined and hence no $f_n(n > 3)$ can be determined.

Specifying the values of f_1 and f_2 is called initial condition. Thus the complete definition of Fibonacci sequence is given by the following recurrence relation and initial conditions.

$$f_n = f_{n-1} + f_{n-2}.$$

$$f_1 = f_2 = 1.$$

The elements of the Fibonocci sequence are $(1, 1, 2, 3, 5, 8, 13, 21, \ldots)$. Note that we can give any values to f_1 and f_2 (not necessarily 1) and correspondingly determine the other elements of the Fibonocci sequence.

Example Consider the recurrence relation $a_n = a_{n-1} + n (n \geq 2)$ with initial condition $a_1 = 1$. Then, $a_2 = a_1 + 2 = 1 + 2 = 3$, $a_3 = a_2 + 3 = 3 + 3 = 6$, $a_4 = a_3 + 4 = 6 + 4 = 10$, etc. Hence the sequence is $(1, 3, 6, 10, \ldots)$. We note that in a recurrence relation, any term a_n after certain stage (in the case of Fibonacci sequence, third term onwards and in the above example, second term onwards) can be expressed in terms of previous terms.

Suppose a recurrence relation is of the form

$$a_n = c_1 a_{n-1} + c_2 a_{n-2} + \cdots + c_r a_{n-r},$$

where $c_1, c_2, \ldots, c_r$ are constants. This recurrence relation is said to be linear, homogeneous of degree r. Solving this recurrence relation means finding an explicit form for a_n independent of its previous terms.

Examples

1. The recurrence relation $a_n = -17.6a_{n-1}$ is linear, homogeneous of degree 1.
2. The recurrence relation for Fibonocci sequence is linear, homogeneous of degree 2.
3. The recurrence relation $a_n = 5a_{n-1} - 6a_{n-2} + 11a_{n-3}$ is linear, homogeneous of degree 3.
4. The recurrence relation $a_n = a_{n-1} + n$ is not linear, not homogeneous.
5. The recurrence relation $a_n = 2^n a_{n-1}$ is not linear, but homogeneous.
6. The recurrence relation $a_n = 5a_{n-1} + 3$ is linear, not homogeneous.

We will now see how to solve a recurrence relation. That is, how to find an explicit form of a_n. The method of backtracking involves repeated substitutions as the following example will show.

Solved problems Using the technique of backtracking, solve the following recurrence relations.

(i) $a_n = 8.3a_{n-1},\ a_1 = -6.$

Solution Repeatedly applying $a_n = 8.3a_{n-1}$, we obtain

$$a_n = 8.3(8.3a_{n-2}) = (8.3)^2 a_{n-2} = (8.3)^3 a_{n-3}.$$

Generalising, $a_n = (8.3)^{n-1} a_{n-(n-1)}$

$$= (8.3)^{n-1} a_1 = -6 \times (8.3)^{n-1}.$$

(ii) $a_n = a_{n-1} + n,\ a_1 = 1.$

Solution Repeatedly applying, $a_n = a_{n-1} + n$, we cbtain

$$a_n = a_{n-2} + n - 1 + n = a_{n-3} + n - 2 + n - 1 + n.$$

$$= a_{n-(n-1)} + 2 + 3 + \cdots + n = a_1 + 2 + 3 + \ldots n$$

$$= 1 + 2 + 3 + \cdots + n = \frac{n(n+1)}{2}.$$

(iii) $a_n = a_{n-1} - 2,\ a_1 = 0.$

Solution Repeatedly applying, $a_n = a_{n-1} - 2$, we obtain

$$a_n = a_{n-2} - 2 - 2 = a_{n-2} - 2 \times 2 = a_{n-3} - 2 - 2 - 2$$

$$= a_{n-3} - 2 \times 3.$$

Generalising,

$$a_n = a_{n-(n-1)} - 2 \times (n-1) = a_1 - 2 \times (n-1) = -2(n-1).$$

Exercise Using the method of backtracking, solve the following recurrence relations.

1. $a_n = 5a_{n-1} + 3,\ a_1 = 2.$
2. $a_n = -1.1a_{n-1},\ a_1 = 5.$

Backtracking (though very easy) is not a very efficient method of solving a recurrence relation. All recurrence relations cannot be solved this way. For solving a linear, homogenous recurrence relation of degree 1, we should be given one initial condition. In general, for solving a linear, homogeneous recurrence relation of degree r, we should be given r initial conditions.

Consider a linear, homogeneous recurrence relation of degree r, say

$$a_n = c_1 a_{n-1} + c_2 a_{n-2} \cdots + c_r a_{n-r}.$$

We can associate with this recurrence relation the equation $x^r = c_1 x^{r-1} + c_2 x^{r-2} + \cdots + c_r$. This equation of degree r is called the characteristic equation associated with the given recurrence relation. We concentrate only on recurrence relations of degree 2. Substituting $r = 2$, the characteristic equation becomes $x^2 = c_1 x + c_2$, which is the same as $x^2 - c_1 x - c_2 = 0$.

Theorem

(i) If the roots of the characteristic equation $x^2 - c_1 x - c_2 = 0$ are x_1 and $x_2 (x_1 \neq x_2)$, then the solution of the recurrence relation $a_n = c_1 a_{n-1} + c_2 a_{n-2}$ is given by $a_n = k_1 x_1^n + k_2 x_2^n$, where k_1 and k_2 are constants determined using the initial conditions.

(ii) If the roots of the characteristic equation $x^2 - c_1 x - c_2 = 0$ are equal, say x_1, then the solution of the recurrence relation $a_n = c_1 a_{n-1} + c_2 a_{n-2}$ is given by $a_n = k_1 x_1^n + k_2 n x_1^n$, where k_1 and k_2 are constants determined using initial conditions.

Proof

(i) Since x_1 and x_2 are roots of the equation $x^2 - c_1 x - c_2 = 0$, we have $x_1^2 - c_1 x_1 - c_2 = 0$ and $x_2^2 - c_1 x_2 - c_2 = 0$. Now,

$$\begin{aligned}
a_n &= k_1 x_1^n + k_2 x_2^n = k_1 x_1^{n-2} x_1^2 + k_2 x_2^{n-2} x_2^2 \\
&= k_1 x_1^{n-2}(c_1 x_1 + c_2) + k_2 x_2^{n-2}(c_1 x_2 + c_2) \\
&= k_1 c_1 x_1^{n-1} + k_1 c_2 x_1^{n-2} + k_2 c_1 x_2^{n-1} + k_2 c_2 x_2^{n-2} \\
&= c_1 (k_1 x_1^{n-1} + k_2 x_2^{n-1}) + c_2 (k_1 x_1^{n-2} + k_2 x_2^{n-2}) \\
&= c_1 a_{n-1} + c_2 a_{n-2}.
\end{aligned}$$

(ii)

$$\begin{aligned}
a_n &= k_1 x_1^n + k_2 n x_1^n \\
&= k_1 x_1^{n-2} x_1^2 + k_2 n x_1^{n-2} x_1^2 \\
&= k_1 x_1^{n-2}(c_1 x_1 + c_2) + k_2 n x_1^{n-2}(c_1 x_1 + c_2) \\
&= k_1 c_1 x_1^{n-1} + k_1 c_2 x_1^{n-2} + k_2 n c_1 x_1^{n-1} + k_2 n c_2 x_1^{n-2} \\
&= c_1 (k_1 x_1^{n-1} + k_2 n x_1^{n-1}) + c_2 (k_1 x_1^{n-2} + k_2 n x_1^{n-2}) \\
&= c_1 a_{n-1} + c_2 a_{n-2}.
\end{aligned}$$

More on recurrence relations can be found in Chapter 12.

Solved problems

Solve the following recurrence relations.

1. $a_n = 4a_{n-1} + 5a_{n-2}$, given that $a_1 = 2$ and $a_2 = 6$.

   Solution The corresponding characteristic equation is $x^2 - 4x - 5 = 0$. The roots of this equation are 5 and -1. Since the roots are distinct, the solution of the given recurrence relation is $a_n = k_1 5^n + k_2(-1)^n$. Substituting $a_1 = 2$ and $a_2 = 6$, we obtain $5k_1 - k_2 = 2$ and $25k_1 + k_2 = 6$. Solving the two equations, we obtain $k_1 = \frac{4}{15}$ and $k_2 = \frac{-2}{3}$. Hence the solution of the given recurrence relation is $a_n = \left(\frac{4}{15}\right) 5^n + \left(\frac{-2}{3}\right)(-1)^n$.

2. $a_n = -6a_{n-1} - 9a_{n-2}$, given that $a_1 = 2.5$ and $a_2 = 4.7$.

   Solution The corresponding characteristic equation is $x^2 + 6x + 9 = 0$. The roots of this equation are -3 and -3. Since the roots are equal, the solution of the given recurrence relation is $a_n = k_1(-3)^n + k_2 n(-3)^n$. Substituting $a_1 = 2.5$ and $a_2 = 4.7$, we obtain $-3k_1 - 3k_2 = 2.5$ and $9k_1 + 18k_2 = 4.7$. Solving the two equations, we obtain $k_1 = -1.82$ and $k_2 = 0.989$. Hence the solution of the given recurrence relation is $a_n = -1.82(-3)^n + 0.989n(-3)^n$.

3. $a_n = 2a_{n-2}$, given that $a_1 = \sqrt{2}$ and $a_2 = 6$.

   Solution The corresponding characteristic equation is $x^2 - 2 = 0$. The roots of this equation are $x = \sqrt{2}$ and $x = -\sqrt{2}$. Since the roots are distinct, the solution of the given recurrence relation is $a_n = k_1(\sqrt{2})^n + k_2(-\sqrt{2})^n$. Substituting $a_1 = \sqrt{2}$ and $a_2 = 6$, we obtain $k_1(\sqrt{2}) + k_2(-\sqrt{2}) = \sqrt{2}$ and $2k_1 + 2k_2 = 6$. Solving the two equations we obtain $k_1 = 2$ and $k_2 = 1$. Hence the solution of the given recurrence relation is $a_n = 2(\sqrt{2})^n + (-\sqrt{2})^n$.

4. $a_n = -3a_{n-1} - 2a_{n-2}$, given that $a_1 = -2$ and $a_2 = 4$.

   Solution The corresponding characteristic equation is $x^2 + 3x + 2 = 0$. The roots of this equation are -2 and -1. Since the roots are distinct, the solution of the given recurrence relation is $a_n = k_1(-2)^n + k_2(-1)^n$. Substituting $a_1 = -2$ and $a_2 = 4$, we obtain $k_1(-2) - k_2 = -2$ and $4k_1 + k_2 = 4$. Solving the two equations we obtain $k_1 = 1$ and $k_2 = 0$. Hence the solution of the given recurrence relation is $a_n = (-2)^n$.

5. $a_n = 4a_{n-1} - 4a_{n-2}$, given that $a_1 = 1$ and $a_2 = 7$.

   Solution The corresponding characteristic equation is $x^2 - 4x + 4 = 0$. The roots of this equation are 2 and 2. Since the roots are equal, the solution of the given recurrence relation is $a_n = k_1 2^n + k_2 n 2^n$. Substituting $a_1 = 1$ and $a_2 = 7$, we obtain $2k_1 + 2k_2 = 1$ and $4k_1 + 8k_2 = 7$. Solving the two equations, we obtain $k_1 = \frac{-3}{4}$ and $k_2 = \frac{5}{4}$. Hence the solution of the given recurrence relation is $a_n = \left(\frac{-3}{4}\right) 2^n + \left(\frac{5}{4}\right) n2^n$.

6. Solve the Fibonacci sequence $f_n = f_{n-1} + f_{n-2},\ f_1 = f_2 = 1$.

Solution The corresponding characteristic equation is $x^2 - x - 1 = 0$. The roots of this equation are $\frac{(1+\sqrt{5})}{2}$ and $\frac{(1-\sqrt{5})}{2}$. Since the roots are distinct, the solution of the given recurrence relation is

$$f_n = k_1\left(\frac{1+\sqrt{5}}{2}\right)^n + k_2\left(\frac{1-\sqrt{5}}{2}\right)^n.$$

Substituting $f_1 = f_2 = 1$, we obtain

$$k_1\left(\frac{1+\sqrt{5}}{2}\right) + k_2\left(\frac{1-\sqrt{5}}{2}\right) = 1$$

and

$$k_1\left(\frac{1+\sqrt{5}}{2}\right)^2 + k_2\left(\frac{1-\sqrt{5}}{2}\right)^2 = 1.$$

Solving the two equations, we obtain $k_1 = \frac{1}{\sqrt{5}}$ and $k_2 = \frac{-1}{\sqrt{5}}$. Hence the solution of the given recurrence relation is

$$f_n = \left(\frac{1}{\sqrt{5}}\right)\left(\frac{1+\sqrt{5}}{2}\right)^n + \left(\frac{-1}{\sqrt{5}}\right)\left(\frac{1-\sqrt{5}}{2}\right)^n.$$

Exercises

Solve the following recurrence relations.

(a) $a_r - 7a_{r-1} + 10a_{r-2} = 0$ given that $a_0 = 0$ and $a_1 = 3$.
(b) $a_r - 4a_{r-1} + 4a_{r-2} = 0$ given that $a_0 = 1$ and $a_1 = 6$.
(c) $a_r + a_{r-1} + a_{r-2} = 0$ given that $a_0 = 0$ and $a_1 = 2$.
(d) $a_r - a_{r-1} - a_{r-2} = 0$ given that $a_0 = 1$ and $a_1 = 1$.
(e) $a_r - 2a_{r-1} + 2a_{r-2} - a_{r-3} = 0$ given that $a_0 = 2$, $a_1 = 1$ and $a_2 = 1$.

6 Relations

6.1 Introduction

We are very familiar with the concept of relations, particularly in the context of relationships among humans in a society. In this chapter we define the mathematical concepts of a relation and an equivalence relation. We also consider the various manipulations that can be performed on relations. We prove the fundamental theorem concerning equivalence relation and partition.

Those who may be wondering what purpose an innocent looking definition like an equivalence relation can serve, may refer to Myhill–Nerode theorem which is basically a theorem about minimizing the number of states in a finite automaton. In this theorem a relation is defined in a very ingenious way, this relation is shown to be an equivalence relation and the corresponding equivalence classes are the states of the minimum state DFA.

6.2 Definition of Relation

All of us are very familiar with the word relation. Some of the relationships that exist between human beings are that of an uncle, aunt, father, mother, etc. Before attempting to give the definition of relation, we will look at some of these familiar relationships from a mathematical viewpoint.

Consider the following sets.

X = {Rama, Raju, Sita, Krishna, Tom, Lakshman}

Y = {Govind, John, Viswam, Usha, Mallikarjuna, Mary, Chandan}

Assume that Rama is the father of Viswam and Govind, Raju is the father of Mallikarjuna, Krishna is the father of Usha and Chandan and Tom is the father of John and Mary. This relation 'father of' can be represented by forming a set of ordered pairs.

$$R = \{(\text{Rama, Govind}), (\text{Rama, Viswam}), (\text{Krishna, Usha}), (\text{Krishna, Chandan}), (\text{Tom, John}), (\text{Tom, Mary}), (\text{Raju, Mallikarjuna})\}.$$

The relation of fatherhood is given the name R. We note that the ordered pair (x, y) belongs to R if and only if x is the father of y. For example, since Tom is the father of John, we have included the ordered pair (Tom, John) in R. Since Krishna is not the father of Govind, the ordered pair (Krishna, Govind) does not belong to R. Since Tom is the father of John, John cannot be the father of Tom. Hence the ordered pair (Tom, John) belongs to R but (John, Tom) does not belong to R.

We note that $X \times Y$ consists of 42 ordered pairs and the relation R is only a subset of $X \times Y$.

Consider two sets A and B. Then a relation R from A to B is nothing but a subset of $A \times B$. If A contains m elements and B contains n elements, then $A \times B$ contains $m \times n$ elements. Hence the number of subsets of $A \times B$ is $2^{m \times n}$. Each of them gives rise to a relation from A to B and hence the number of relations from A to B is $2^{m \times n}$. If R is a relation from A to B, then any element of R is of the form (a, b), where a belongs to A and b belongs to B. If (a, b) belongs to R, we say that a is related to b through R and denote it by writing aRb. If (a, b) does not belong to R, then we say that a is not related to b through R and we denote it by writing $a \not R b$.

Example

Let $A = \{6, 13, 28, 42\}$ and $B = \{12, 20, 21, 58, 83\}$.
Let $R = \{(13, 12), (28, 12), (42, 12), (28, 20), (42, 20), (42, 21), (28, 21)\}$.

Then R is a relation from A to B. We know that if a belongs to A and b belongs to B, then (a, b) belongs to R if and only if $a > b$.

Even though a relation can be defined from any set A to any set B, most of the times we will be concentrating on relations from A to itself. A relation R from A to A which is a subset of $A \times A$ will be called a relation on A. These relations possess very interesting properties. Unless otherwise specified, the relations we consider hereafter will all be relations on a set A.

6.3 Types of Relations

Reflexive relation

Let R be a relation on a set A. R is said to be *reflexive* if every element of A is related to itself through the relation R. R is reflexive if and only if (x, x) belongs to R for all x belonging to A.

Examples

1. Let $X = \{$Rama, Raju, Sita, Krishna, Tom, Lakshman$\}$. Define a relation R on X as xRy if and only if x is the brother of y. This relation is not reflexive since (Sita, Sita) does not belong to R.
2. Let $A = \{6, 13, 28\}$ and let $S = \{(6, 13), (6, 28), (13, 6), (13, 28), (28, 6), (28, 13)\}$. S is not a reflexive relation on A.

Note If we want to prove that a relation R on a set A is reflexive, we have to prove that for all x belonging to A, (x, x) belongs to R. To prove that R is not reflexive, we have to find at least one element y such that (y, y) does not belong to R.

Irreflexive relation

We say that a relation R on a set A is *irreflexive* if (x, x) does not belong to R for all x belonging to A. The relation S in Example 2 given above is irreflexive.

Note If a relation R is irreflexive, then R is not reflexive. However, if R is not reflexive, we cannot say R is irreflexive as the following example shows.

Let $F = \{$mango, chickoo, grape$\}$ and let

$R = \{$(mango, mango), (mango, grape), (chickoo, grape), (chickoo, chickoo)$\}$.

R is not reflexive since (grape, grape) does not belong to R. R is also not irreflexive since (mango, mango) belongs to R.

Problem If a set A contains n elements, can you tell the number of relations on A which are reflexive?

Symmetric relation

Definition A relation R on a set A is said to be *symmetric* if whenever (x, y) belongs to R, (y, x) also belongs to R.

As the name itself implies, in a symmetric relation there should be symmetry. Whenever xRy, then yRx. Note that we want yRx only when xRy. We do not bother to see whether yRx if x is not related to y.

Examples

1. Consider the relation "brother of" given earlier. This relation is not symmetric. If x is a brother of y, then y may be a sister of x.
2. Let $A = \{18, 13, 27, 42\}$ and let $R = \{(18, 13), (18, 42), (20, 13), (13, 18), (42, 18), (13, 20)\}$.

R is symmetric and also irreflexive and hence not reflexive. Note that since $(27, 42)$ does not belong to R, we do not bother to check whether $(42, 27)$ belongs to R.

Problem If a set A contains n elements, can you tell the number of relations on A which are symmetric?

Note If we want to prove that a relation R is symmetric, we have to prove that (y, x) belongs to R whenever (x, y) belongs to R. To prove that R is not symmetric we have to find two elements a, b such that (a, b) belongs to R but (b, a) does not belong to R.

Asymmetric relation

A relation R is said to be *asymmetric* (opposite of symmetric) if whenever (x, y) belongs to R, (y, x) does not belong to R.

We note that a relation which is symmetric cannot be asymmetric. If a relation is not symmetric we cannot say it is asymmetric. Also note that in an asymmetric relation R on a set A, (a, a) cannot belong to R for any $a \in A$. Hence an asymmetric relation is irreflexive.

Example Let $A = \{6, 13, 28, 42\}$ and let $R = \{(6, 6), (6, 28), (28, 6), (13, 13), (13, 42), (28, 28), (28, 42), (42, 42)\}$. R is neither symmetric nor asymmetric. However, R is reflexive.

Antisymmetric relation

A relation R on a set A is said to be *antisymmetric* if whenever (x, y) belongs to R and (y, x) belongs to R, then $x = y$. In other words, if x is not equal to y, then both (x, y) belongs to R and (y, x) belongs to R cannot happen. In an antisymmetric relation, the only way by which both xRy and yRx are possible is when $x = y$.

Examples

1. Consider the sets A and R given above. Then R is not antisymmetric.
2. Let $S = \{(6, 6), (6, 28), (13, 13), (13, 42), (28, 28), (28, 42), (42, 42)\}$. Then S is an antisymmetric relation on A. S is not asymmetric since $(6, 6) \in S$.
3. Let $M = \{\text{Rama, Krishna, Uma}\}$ and let $R = \{(\text{Rama, Krishna})\}$. Then R is not reflexive; it is irreflexive, not symmetric; it is asymmetric and antisymmetric.

Note that an antisymmetric relation need not be an asymmetric relation (consider the relation S given above) whereas an asymmetric relation is antisymmetric.

Transitive relation

A relation R on a set A is said to be *transitive* if whenever (x, y) belongs to R and (y, z) belongs to R, then (x, z) belongs to R, i.e., whenever xRy and yRz, then xRz.

Problem If a set A contains n elements, can you tell the number of relations on A which are transitive?

Examples

1. Consider the relation 'brother of' given earlier. If x is a brother of y and y is a brother of z, then x is a brother of z. Hence the relation 'brother of' is a transitive relation.
2. Let $A = \{\text{tiger, camel, lion, cat}\}$ and let $R = \{(\text{tiger, camel}), (\text{camel, lion}), (\text{lion, lion}), (\text{tiger, cat}), (\text{tiger, lion})\}$. R is transitive and antisymmetric. R is not reflexive, not irreflexive, not symmetric and not asymmetric.
3. Let $B = \{\text{crow, peacock, parrot}\}$ and let $R = \{(\text{crow, parrot}), (\text{peacock, parrot}), (\text{crow, crow}), (\text{peacock, peacock}), (\text{parrot, parrot})\}$. R is reflexive, antisymmetric and transitive. R is not symmetric and not asymmetric. R is an example of a partial order relation.
4. Consider the set B given above and let $T = \{(\text{crow, peacock}), (\text{crow, crow}), (\text{peacock, parrot}), (\text{parrot, parrot})\}$. T is not transitive because (crow, peacock) belongs to T and (peacock, parrot) belongs to T but (crow, parrot) does not belong to T.

A relation R on a set A is said to be an *equivalence relation* if R is reflexive, symmetric and transitive. A relation R on a set A is said to be a *partial order relation* if R is reflexive, antisymmetric and transitive.

Examples

1. Consider the relation '=' defined on N. That is, xRy if and only if $x = y$. Then $R = \{(1,1), (2,2), (3,3), (4,4), \ldots\}$, which is an equivalence relation.
2. Consider the relation '$\leq$' defined on N. That is, xRy if and only if $x \leq y$. Then R is a partial order relation.

Note A partial order relation on a set A is normally denoted by $\leq$. This is only a symbol and has nothing to do with the usual $\leq$ symbol between numbers.

If a set A has a partial order relation $\leq$, then $(A, \leq)$ is called a partially ordered set (or) simply a poset.

A partial order relation $\leq$ satisfies the following.

1. $x \leq x$ for every x (reflexive).
2. If $x \leq y$ and $y \leq x$, then $x = y$ (antisymmetric).
3. If $x \leq y$ and $y \leq z$, then $x \leq z$ (transitive).

Consider a poset $(A, \leq)$. If $a, b \in A$, then a and b are said to be comparable if either $a \leq b$ or $b \leq a$.

Note In a poset $(A, \leq)$, there can exist elements x and y such that neither $x \leq y$ nor $y \leq x$ holds. This means that it is not necessary that any two elements of a poset be comparable.

Definition A poset $(A, \leq)$ is called a *linearly ordered set* or a *totally ordered set* or a *chain* if any two elements of A are comparable. This means for any two elements $x, y \in A$, either $x \leq y$ or $y \leq x$.

Example Consider the usual less than or equal to relation on N. Then N is a totally ordered set with this relation. For any two elements $m, n \in N$, either $m \leq n$ or $n \leq m$.

Solved problems

1. Show that if a relation R on a set A is transitive and irreflexive, then R is asymmetric.

   Solution Suppose xRy. To prove R is asymmetric, we have to prove that $y \not R x$. Let, if possible, yRx. Since xRy and yRx and R is transitive, it follows that xRx and hence R is not irreflexive (by definition of irreflexive relation). This is a contradiction to the fact that R is irreflexive. Hence $y \not R x$ proving that R is asymmetric.

2. Let R be a non-empty relation on a set A. Suppose R is symmetric and transitive. Prove that R is not irreflexive.

Solution Since R is non-empty, R contains at least one element (x, y). Since R is symmetric, $(y, x) \in R$. Since R is transitive, $(x, y) \in R$ and $(y, x) \in R$ imply $(x, x) \in R$. This proves that R is not irrefllexive.

3. A relation R on a set A is said to be circular if aRb and bRc imply that cRa. Show that R is reflexive and circular if and only if R is an equivalence relation.

Solution First assume that R is an equivalence relation. We will show that R is reflexive and circular. By definition of equivalence relation, R is reflexive. To prove R is circular, assume that aRb and bRc. Since R is an equivalence relation, R is transitive and hence aRc. Again R is symmetric (since R is an equivalence relation) implying cRa. This proves that R is circular.

Conversely, assume that R is reflexive and circular. We will prove that R is an equivalence relation. For this we have to prove that R is reflexive, symmetric and transitive. It is given that R is reflexive. To prove R is symmetric, assume that xRy. Since R is given to be reflexive, xRx. Since xRx and xRy, by definition of circular relation yRx and hence R is symmetric. To prove R is transitive, assume that xRy and yRz. Since R is given to be circular, we have zRx. Since we have already proved that R is symmetric, it follows that xRz and hence R is transitive. Since R is reflexive, symmetric and transitive, R is an equivalence relation.

Definition Let R be a relation from A to B. Then R is simply a subset of $A \times B$ and hence all the elements of A need not be related to elements of B. It is also not necessary that elements of A be related to all the elements of B. This means some elements of B may be left out from the ordered pairs in R. The set of all elements of A which are related to elements of B is a subset of A called the *domain of* R and is denoted by Dom (R). The set of all elements of B for which there exist elements in A which are related to those elements in B is a subset of B called the *range of* R and is denoted by Ran (R). We have Dom $(R) \subseteq A$ and Ran $(R) \subseteq B$.

Example Let $A = \{1, 2, 3, 4\}$, $B = \{a, b, c\}$ and let $R = \{(1, a), (1, c), (2, b), (4, a)\}$. Then Dom $(R) = \{1, 2, 4\}$ and Ran $(R) = \{a, b, c\}$. We note that Dom (R) is a proper subset of A while Ran $(R) = B$. In general, Dom (R) and Ran (R) can be proper subsets of A and B respectively.

6.4 Representation of Relations

(a) Matrix representation of a relation

Consider a relation R from a set A to a set B. We can associate a matrix with R as follows.

The elements of A correspond to the rows of the matrix and the elements of B correspond to the columns of the matrix. If $(a, b) \in R$ where $a \in A$ and $b \in B$, then the element in the matrix corresponding to row a and column b is made 1. Otherwise this entry is made zero. The resulting matrix is the matrix associated with the relation R and is denoted by M_R.

Example The matrix corresponding to the above relation R is $\begin{pmatrix} 1 & 0 & 1 \\ 0 & 1 & 0 \\ 0 & 0 & 0 \\ 1 & 0 & 0 \end{pmatrix}$.

Note

1. If A contains m elements and B contains n elements, then the matrix M_R is an $m \times n$ matrix.
2. Suppose R is a relation on a set A which contains n elements. The corresponding matrix is a square matrix of order n. If the relation R is reflexive, then all diagonal elements of this matrix are 1s. If R is irreflexive, then all diagonal elements are 0s. If R is symmetric, then the matrix M_R is a symmetric matrix.

(b) Digraph representation of a relation

Consider a relation R on a set A. We can associate a digraph (directed graph) with R whose construction is given below.

Think of the elements of A as representing the vertices. Suppose $(a, b) \in R$, where $a, b \in A$. We know that a and b are vertices. Draw a directed edge from a to b. Repeat this for each ordered pair belonging to R. The resulting figure is a directed graph (digraph) associated with the relation R. Simply speaking, the digraph associated with a relation R on a set A has all its vertices representing the elements of A and a directed edge from x to y if and only if $(x, y) \in R$.

Example Let $A = \{$cat, tiger, lion, dog$\}$ and let $R = \{$(cat, tiger), (tiger, dog), (lion, lion), (cat, dog)$\}$. Then R is represented by the following digraph.

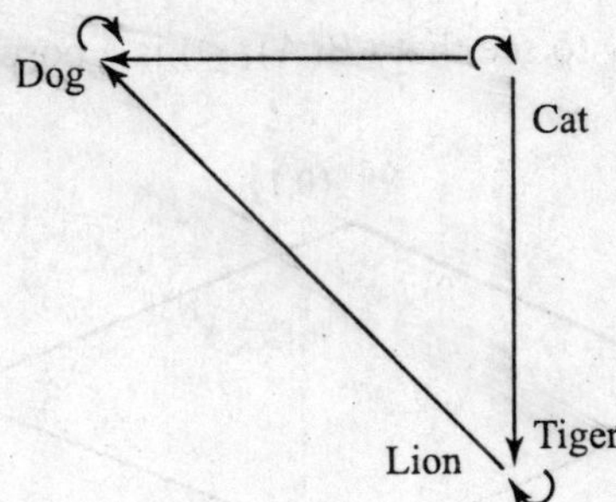

Note that there is a loop (also called a cycle of length 1) at vertex lion since (lion, lion) $\in R$. If the relation R is reflexive, then there will be a loop at every vertex. Suppose $(x, y) \in R$ and $(y, x) \in R$. Then there is a directed edge from x to y and also a directed edge from y to x. If the relation R is symmetric, then whenever there is an edge from vertex a to vertex b, there is also an edge from vertex b to vertex a. We can replace these two directed edges by a single undirected edge connecting vertices a and b. This edge can be used to go from vertex a to vertex b and also from vertex b to vertex a. We thus obtain an undirected graph, i.e., a graph in which no edge has a direction. Suppose $(x, y) \in R, (y, z) \in R$ and $(x, z) \in R$. Then there is a directed edge from vertex x to vertex y, a directed edge from vertex y to vertex z and a directed edge from vertex x to vertex z. In particular, if the relation R is transitive, then whenever there is a directed edge from vertex a to vertex b and a directed edge from vertex b to vertex c, there is a directed edge from vertex a to vertex c.

Now consider a relation R on a set A which is a partial order relation. We know that R is reflexive and hence in the corresponding digraph there is a loop at every vertex. We can safely drop all these loops with a clear understanding that a loop exists at every vertex. Since R is anti-symmetric, if there is an edge from vertex a to vertex b, there cannot be an edge from vertex b to vertex a. In this case we agree to write b above a and draw a line connecting a and b. It is understood that edges are in upward directions and hence we can safely omit the directions from all the edges. Since R is transitive, we know that whenever there is an edge from vertex x to vertex y and an edge from vertex y to vertex z, then there is an edge from vertex x to vertex z. We can safely omit this edge from vertex x to vertex z with a clear understanding that this edge does exist. This simplified figure is called the *Hasse diagram* associated with the partial order relation R.

Examples

(a) Let $A = \{0, 1\}$ and consider $P(A)$ the power set of A. If $X, Y \in P(A)$ (we know that X and Y are subsets of A), define $X \leq Y$ if and only if

$X \subseteq Y$. It is easy to see that $(P(A), \leq)$ is a poset. Its Hasse diagram is shown below.

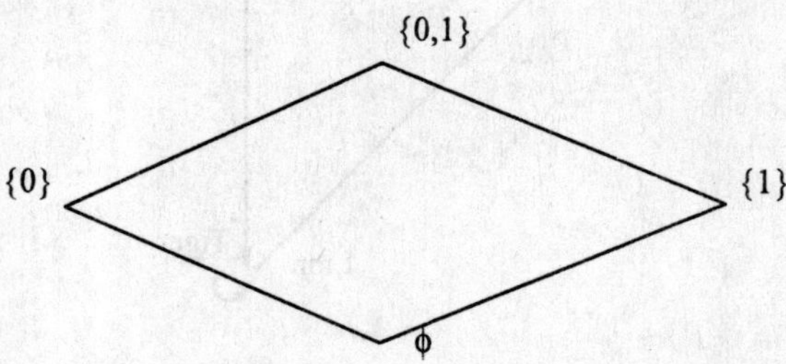

Note The sets {0} and {1} are not comparable. Neither of them is a subset of the other. Hence they are at the same level in the Hasse diagram. Since {1} is a subset of {0, 1}, we have $\{1\} \leq \{0, 1\}$. Hence {0, 1} is at a level higher than {1} and we have an edge from {1} to {0, 1}. Even though $\phi \subseteq \{0, 1\}$, we have omitted the edge from ϕ to {0, 1} since there is an edge from ϕ to {0} and an edge from {0} to {0, 1}. Since the edges are assumed to be in the upward direction, we have omitted directions from all the edges. We have also omitted loops from all the vertices.

(b) Let $D(81)$ denote the set of all positive divisors of 81. $D(81) = \{1, 3, 9, 27, 81\}$. If $a, b \in D(81)$, define $a \leq b$ if and only if a divides b (this means that when b is divided by a, the remainder is 0). Then $(D(81), \leq)$ is a partial order relation. It is easy to see that $\leq$ is reflexive, antisymmetric and transitive. Its Hasse diagram is as shown below.

It can be seen that $(D(81), \leq)$ is a totally ordered set.

(c) Consider $P(S)$ where S is any set. Then $(P(S), \subseteq)$ is a partially ordered set which is not a totally ordered set. For any two elements a and b belonging to S, $\{a\} \in P(S)$ and $\{b\} \in P(S)$, but neither is $\{a\}$ a subset of $\{b\}$ nor $\{b\}$ a subset of $\{a\}$.

6.5 Compatibility Relation

Definition A relation R on a set S is said to be a *compatibility relation* if R is reflexive and symmetric. Clearly, an equivalence relation is a compatibility relation but a compatibility relation need not be an equivalence relation as the following example shows.

Example Let $S = \{X_1, X_2, X_3, X_4, X_5\}$ where X_1 = tiger. X_2 = goat, X_3 = ass, X_4 = sheep and X_5 = cat. Define a relation R on S as $X_i R X_j$ if and only if X_i andX_j have at least one common letter. Clearly, R is reflexive and symmetric and hence R is a compatibility relation. However, R is not transitive since $(X_1, X_2) \in R$ and $(X_2, X_3) \in R$, but $(X_1, X_3) \notin R$. (Note that 'tiger' and 'ass' do not have any common letters.) Hence R is not an equivalence relation.

Since a compatibility relation is reflexive, all the diagonal entries are 1s in its matrix representation and there is a loop in the digraph representation at every vertex. Since a compatibility relation is symmetric, the corresponding matrix is a symmetric matrix. In the digraph representation, if there is an edge from one vertex to another vertex, there should be an edge in the opposite direction. In view of these facts, in the matrix representation of a compatibility relation we will write only the elements which are below the main diagonal (we can get other elements easily), and in the digraph representation we will drop all loops and draw an undirected edge connecting X_i and X_j if $X_i R X_j$.

The matrix representation of the above relation R is as follows.

X_2	1			
X_3	0	1		
X_4	1	0	1	
X_5	1	1	1	0
	X_1	X_2	X_3	X_4

The graphical representation of the above relation R is shown below.

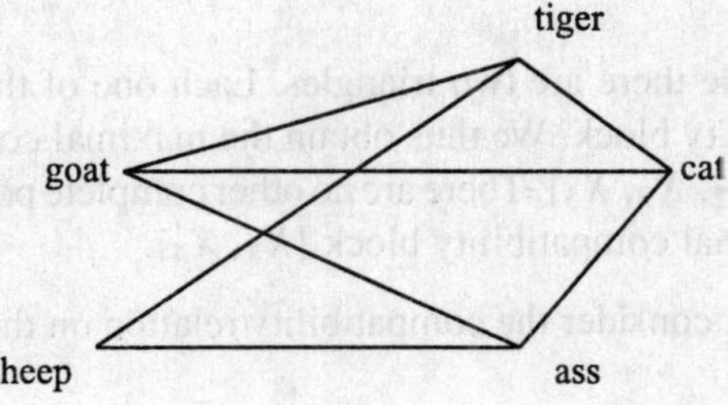

Definition Let R be a compatibility relation on a set S. A subset A of S is said to be a *maximal compatibility block* if every element of A is related to every element of A through R and any element not in A is not related to at least one element of A.

Example Consider the example given above. $A = \{X_1, X_2, X_5\}$ is a maximal compatibility block. Note that every element of A is related to every element of A through R. Also, $(X_1, X_3) \notin R$ and $(X_4, X_5) \notin R$. $\{X_3, X_4\}$ is another maximal compatibility block and $\{X_2, X_3, X_5\}$ is a third maximal compatibility block. We note that the union of these maximal compatibility blocks is S itself. But these maximal compatibility blocks do not form a partition of S since these maximal compatibility blocks are not pairwise disjoint.

Let R be a compatibility relation on a set S. The maximal compatibility blocks can be obtained using the following steps.

Step 1: First draw the graph of the compatibility relation as mentioned above. We do not use the word digraph since the edges do not have any direction.

Step 2: Suppose there is an element X_i in S such that no element other than X_i is related to X_i. (X_i should be related to X_i since R is reflexive). Then $\{X_i\}$ is a maximal compatibility block. Identify all such maximal compatibility blocks each consisting of only one element.

Step 3: Suppose there are two elements X_i and X_j which are compatible but any other element of S is not compatible to either X_i or X_j. Then $\{X_i, X_j\}$ is a maximal compatibility block. Identify all such maximal compatibility blocks each consisting of two elements.

Step 4: From the graph, identify the largest complete polygons. A complete polygon is a polygon in which any two vertices are connected by an edge. A triangle is obviously a complete polygon. Each of these complete polygons will give rise to a maximal compatibility block.

In the above example there are two triangles. Each one of them gives rise to a maximal compatibility block. We thus obtain the maximal compatibility blocks $\{X_1, X_2, X_5\}$ and $\{X_2, X_3, X_5\}$. There are no other complete polygons. By Step 3, we obtain the maximal compatibility block $\{X_3, X_4\}$.

As another example, consider the compatibility relation on the set $\{1, 2, 3, 4, 5\}$ whose matrix is:

2	1			
3	1	1		
5	0	0	1	
6	1	0	1	1
	1	2	3	5

Note that we have omitted 4 from the matrix, which indicates that no other element is related to 4. Its graphical representation is given below.

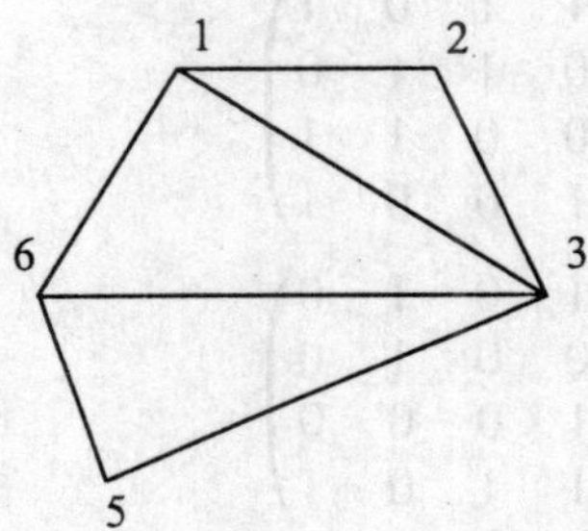

From the graph we can see that the maximal compatibility blocks are {1, 2, 3}, {1, 3, 6}, {3, 5, 6} and {4}. We have included {4} as a maximal compatibility block by Step 2. There is a quadrilateral whose vertices are 1, 2, 3 and 6 but it is not complete since 2 and 6 are not connected.

Exercises

1. Let $A = \{1, 2, 3, 4\}$ and $R = \{(1, 1), (1, 2), (2, 2), (2, 4), (1, 3), (3, 3), (3, 4), (1, 4), (4, 4)\}$. Determine the Hasse diagram of R.
2. In the following problems, consider the partial order of divisibility on the given set A. In each problem, draw the Hasse diagram of the poset and determine whether the poset is linearly ordered.

   (i) $A = \{1, 2, 3, 5, 6, 10, 15, 30\}$.
   (ii) $A = \{2, 4, 8, 16, 32\}$.
   (iii) $A = \{3, 6, 12, 36, 72\}$.
   (iv) $A = \{1, 2, 3, 4, 5, 6, 10, 12, 15, 30, 60\}$.

3. In each of the following problems find the domain, the range, the matrix, and when $A = B$, the digraph of the given relation.

   (i) $A = \{a, b, c, d\}$, $B = \{1, 2, 3\}$ and $R = \{(a, 1), (a, 2), (b, 1), (c, 2), (d, 1)\}$.
   (ii) $A = \{1, 2, 3, 4\}$, $B = \{1, 4, 6, 8, 9\}$ and R is defined as $a\ R\ b$ if and only if $b = a^2$.

(iii) $A = \{1, 2, 3, 4, 8\}$, $B = \{1, 4, 6, 9\}$ and R is defined as $a\ R\ b$ if and only if a divides b.

(iv) $A = \{1, 2, 3, 4, 6\} = B$ and R is defined as $a\ R\ b$ if and only if a is a multiple of b.

(v) $A = \{1, 2, 3, 4, 5\} = B$ and R is defined as $a\ R\ b$ if and only if $a \leq b$.

4. In each of the following problems, determine the relation R on $A = \{a, b, c, d\}$ determined by the matrix

(i) $$M_R = \begin{pmatrix} 1 & 1 & 0 & 1 \\ 0 & 1 & 1 & 0 \\ 0 & 0 & 1 & 1 \\ 1 & 0 & 0 & 1 \end{pmatrix}.$$

(ii) $$M_R = \begin{pmatrix} 1 & 0 & 1 & 0 \\ 0 & 0 & 1 & 0 \\ 1 & 0 & 0 & 0 \\ 1 & 1 & 0 & 1 \end{pmatrix}.$$

5. Find the relation R determined by each of the following digraphs.

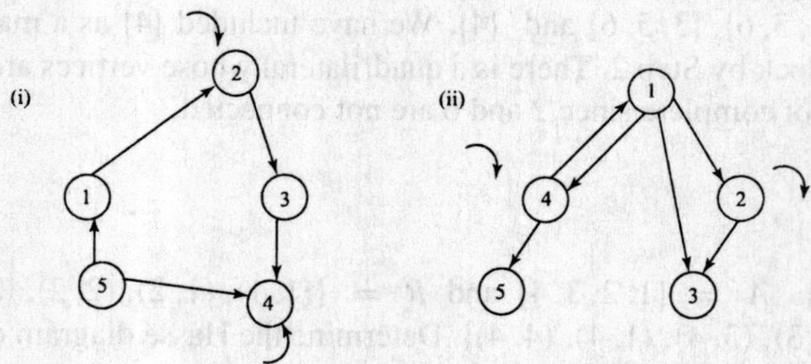

Also, find the matrix M_R in each case.

6. Let $A = \{a, b, c, d, e\}$. Determine the digraph of the relation R on A corresponding to the following matrices.

(i) $$M_R = \begin{pmatrix} 1 & 1 & 0 & 0 & 0 \\ 0 & 0 & 1 & 1 & 0 \\ 0 & 0 & 0 & 1 & 1 \\ 0 & 1 & 1 & 0 & 0 \\ 1 & 0 & 0 & 0 & 0 \end{pmatrix}$$

(ii) $$M_R = \begin{pmatrix} 0 & 0 & 0 & 0 & 1 \\ 0 & 1 & 0 & 0 & 0 \\ 0 & 0 & 1 & 0 & 0 \\ 0 & 0 & 0 & 1 & 0 \\ 1 & 1 & 1 & 0 & 1 \end{pmatrix}$$

7. Let $A = \{1, 2, 3, 4\}$. Determine whether the following relations are reflexive, irreflexive, symmetric, asymmetric, antisymmetric or transitive.

(a) $R = \{(1, 1), (1, 2), (2, 1), (2, 2), (3, 3), (3, 4), (4, 3), (4, 4)\}$.

(b) $R = \{(1, 2), (1, 3), (1, 4), (2, 3), (2, 4), (3, 4)\}$.

(c) $R = \{(1, 2), (1, 3), (3, 1), (1, 1), (3, 3), (3, 2), (1, 4), (4, 2), (3, 4)\}$.

8. Let $A = \{1, 2, 3, 4, 5\}$. Determine whether the relation R whose digraph is given below is reflexive, irreflexive, symmetric, asymmetric, antisymmetric or transitive.

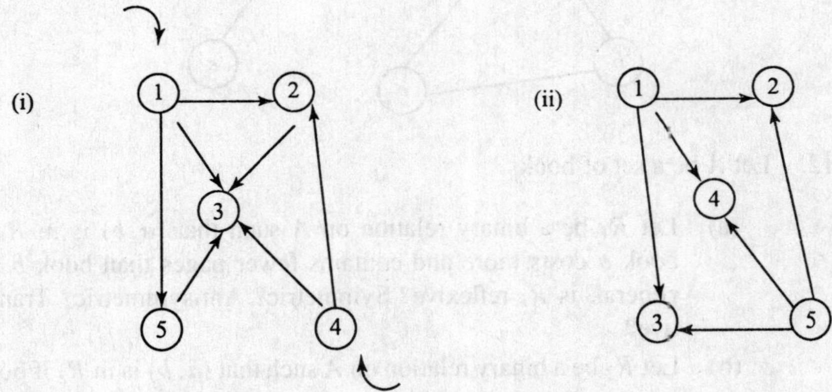

9. Let $A = \{1, 2, 3, 4\}$. Determine whether the relation R whose matrix is given below is reflexive, irreflexive, symmetric, asymmetric, antisymmetric or transitive.

(a) $M_R = \begin{pmatrix} 0 & 1 & 0 & 1 \\ 1 & 0 & 1 & 1 \\ 0 & 1 & 0 & 0 \\ 1 & 1 & 0 & 0 \end{pmatrix}$

(b) $M_R = \begin{pmatrix} 1 & 1 & 0 & 0 \\ 1 & 1 & 0 & 0 \\ 0 & 0 & 1 & 0 \\ 0 & 0 & 0 & 1 \end{pmatrix}$

10. Determine whether the following relations defined on A are reflexive, irreflexive, symmetric, asymmetric, antisymmetric or transitive.

(a) $A = Z$, $a\ R\ b$ if and only if $a \leq b + 1$.

(b) $A = Z^+$, $a\ R\ b$ if and only if $|a - b| \leq 2$. (Z^+ denotes the set of all positive integers).

(c) $A = Z, a\ R\ b$ if and only if $a + b$ is even.

(d) $A = \mathbb{R}$, the set of all real numbers. Define $a\ R\ b$ if and only if $a^2 + b^2 = 4$.

(e) $A = \mathbb{R} \times \mathbb{R}$, $(a, b)R(c, d)$ if and only if $a = c$.

(f) $A = Z^+, a\ R\ b$ if and only if $\text{GCD}(a, b) = 1$.

11. The graph of a symmetric relation R on $A = \{a, b, c, d, e\}$ is given in the following figure. Determine the relation R.

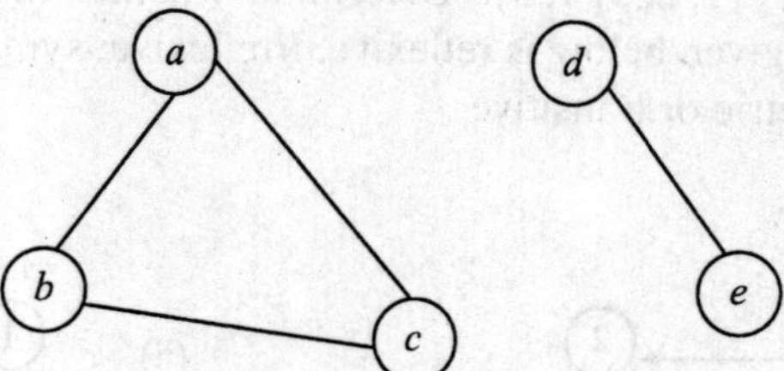

12. Let A be a set of books.

(a) Let R_1 be a binary relation on A such that (a, b) is in R_1 if book a costs more and contains fewer pages than book b. In general, is R_1 reflexive? Symmetric? Antisymmetric? Transitive?

(b) Let R_2 be a binary relation on A such that (a, b) is in R_2 if book a costs more or contains fewer pages than book b. In general, is R_2 reflexive? Symmetric? Antisymmetric? Transitive?

13. Let R be a binary relation on the set of all positive integers such that $R = \{(a, b) | a - b \text{ is an odd positive integer}\}$.

(a) Is R reflexive? Symmetric? Antisymmetric? Transitive? An equivalence relation? A partial order relation?

(b) Repeat part (a) if $R = \{(a, b) | a = b^2\}$.

14. Let R be a binary relation on the set of all strings of 0s and 1s such that $R = \{(a, b) | a \text{ and } b \text{ are strings that have the same number of 0 s}\}$. Is R reflexive? Symmetric? Antisymmetric? Transitive? An equivalence relation? A partial order relation?

15. Let $S = \{1, 2, 3, 4\}$. Consider a relation R on S defined as $R = \{(1, 2), (4, 3), (2, 2), (2, 1), (3, 1)\}$. Show that R is not transitive. Find a relation $R_1 \supseteq R$ such that R_1 is transitive. Can you find another relation $R_2 \supseteq R$ which is also transitive?

16. Let the compatibility relation on a set $\{X_1, X_2, X_3, X_4, X_5, X_6\}$ be given by the following matrix.

X_2	1				
X_3	1	1			
X_4	0	0	1		
X_5	0	0	1	1	
X_6	1	0	1	0	1
	X_1	X_2	X_3	X_4	X_5

Draw the graph and find the maximal compatibility blocks of the relation.

17. Let R be a symmetric and transitive relation on a set A. Show that if for every a in A there exists b in A such that (a, b) is in R, then R is an equivalence relation.
18. Let R be a transitive and reflexive relation on A. Let T be a relation on A such that (a, b) is in T if and only if both (a, b) and (b, a) are in R. Show that T is an equivalence relation.
19. Let R be a binary relation. Let $S = \{(a, b) | (a, c) \in R$ and $(c, b) \in R$ for some $c\}$. Show that if R is an equivalence relation, then S is also an equivalence relation.
20. Let R be a reflexive relation on a set A. Show that R is an equivalence relation if and only if (a, b) and (a, c) are in R implies that (b, c) is in R.

Theorem Let A be a non-empty set and let R be an equivalence relation on A. Then R gives rise to a partition of A.

Proof Since A is non-empty, we can find an element $a \in A$. Define $[a] = \{x \in A | aRx\}$. In other words, $[a]$ is the set of all those elements of A to which a is related. $[a]$ is called the equivalence class of A corresponding to the element a. We find equivalence classes corresponding to each element of A. We will prove that $\{[a] | a \in A\}$ forms a partition of A. For this we have to prove the following.

1. $[a] \neq \phi \ \forall a \in A$.
2. For any two equivalence classes $[a]$ and $[b]$, either $[a] = [b]$ or $[a] \cap [b] = \phi$.
3. $\cup[a] = A$ where the union is taken over all elements $a \in A$.

Since R is reflexive, for every element $a \in A$, we have aRa and hence $a \in [a]$ This proves that $[a] \neq \phi$.

To prove 2, we assume that $[a] \cap [b] \neq \phi$ and we will prove that $[a] = [b]$. Since $[a] \cap [b] \neq \phi$, there is an element $c \in [a] \cap [b]$. Since $c \in [a]$, we have aRc. Since $c \in [b]$, we have bRc. Since R is symmetric, bRc implies cRb. Since R is transitive, aRc and cRb imply aRb which in turn implies bRa since R is symmetric. Now let $x \in [a]$. By definition of equivalence class of a, we have aRx. Since R is transitive, bRa and aRx imply bRx. Hence by definition of

$[b]$, $x \in [b]$. This proves that $[a] \subseteq [b]$. Similarly we can prove that $[b] \subseteq [a]$. Hence $[a] = [b]$.

We now come to the proof of 3.

Since $[a] \subseteq A\ \forall\ a \in A$, it follows that $\cup[a] \subseteq A$. If $d \in A$, then $d \in [d]$ and hence $d \in \cup[a]$. This prove that $A \subseteq \cup[a]$. Hence $\cup[a] = A$.

Theorem Let A be a non-empty set and let P be a partition of A. Then P defines an equivalence relation on A.

Proof Define a relation R as xRy if and only if x and y belong to the same block. We will prove that R is an equivalence relation.

For any element x, since x and x belong to the same block, by definition of R, xRx and hence R is reflexive.

If xRy, then x and y are in the same block which is the same as saying that y and x are in the same block. Hence yRx proving that R is symmetric.

If xRy and yRz, then by definition of R, x and y are in the same block and y and z are in the same block. Hence x and z are in the same block so that xRz. This proves that R is transitive.

Exercises

1. Let $A = \{a, b, c\}$. In each of the following problems, determine whether the relation R, whose matrix M_R is given is an equivalence relation:

   (a) $M_R = \begin{bmatrix} 1 & 0 & 1 \\ 0 & 1 & 0 \\ 0 & 0 & 1 \end{bmatrix}$

   (b) $M_R = \begin{bmatrix} 1 & 0 & 0 \\ 0 & 1 & 1 \\ 0 & 1 & 1 \end{bmatrix}$

2. Determine whether the relation R whose digraph is given below is an equivalence relation.

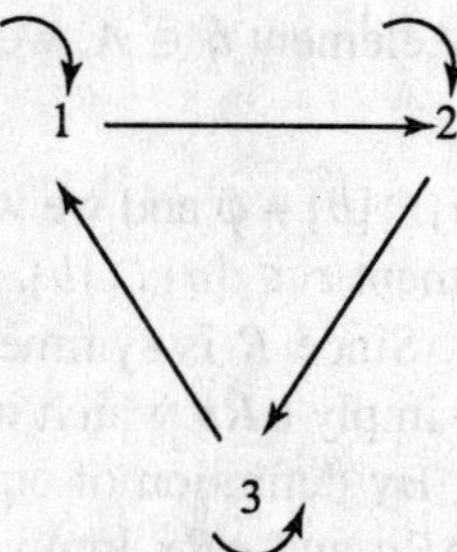

3. Let $A = \{a, b, c, d\}$ and let $R = \{(a, a), (b, a), (b, b), (c, c), (d, d), (d, c)\}$. Determine whether R is an equivalence relation.
4. Let $A = \{1, 2, 3, 4\}$ and let $R = \{(1, 1), (1, 2), (2, 2), (3, 1), (3, 3), (1, 3), (4, 1), (4, 4)\}$. Determine whether R is an equivalence relation.
5. Let $S = \{1, 2, 3, 4, 5\}$ and let $A = S \times S$. Define a relation R on A as $(a, b)R(a_1, b_1)$ if and only if $ab_1 = a_1 b$. Prove that R is an equivalence relation. Also, compute A/R.
6. Let $S = \{1, 2, 3, 4\}$, $A = S \times S$. Define a relation R on A as $(a, b)R(a_1, b_1)$ if and only if $a + b = a_1 + b_1$. Prove that R is an equivalence relation. Compute A/R.
7. If $\{\{1, 3, 5\}, \{2, 4\}\}$ is a partition of $A = \{1, 2, 3, 4, 5\}$, determine the corresponding equivalence relation R.
8. If $\{\{a, c, e\}, \{b, d, f\}\}$ is a partition of $A = \{a, b, c, d, e, f\}$, determine the corresponding equivalence relation R.

If R is an equivalence relation on A, then we define $A/R = \{[x]/x \in A\}$. It is called the quotient set.

Let m be a positive integer and let $R = \{(x, y)|x, y \in Z$ and m divides $x - y\}$. We will prove that R is an equivalence relation. Clearly, m divides $x - x = 0$ so that $(x, x) \in R$. Hence R is reflexive. Suppose $(x, y) \in R$. Then m divides $x - y$, which means $x - y = km (k \in Z)$. Then $y - x = (-k)m$, which means m divides $y - x$ so that $(y, x) \in R$. Hence R is symmetric. Suppose $(x, y) \in R$ and $(y, z) \in R$. Then m divides $x - y$ and m divides $y - z$ so that $x - y = km$ and $y - z = lm \quad (k, l \in Z)$. Adding, we obtain $x - z = (x - y) + (y - z) = km + lm = (k + l)m$. Hence m divides $x - z$ so that $(x, z) \in R$. This proves that R is transitive.

We will now take $m = 3$ and determine the equivalence classes of R.

We have

$$\begin{aligned}
[0] &= \{x \in Z/x \; R \; 0\} \\
&= \{x \in Z/3 \text{ divides } x - 0\} \\
&= \{3k/k \in Z\} \\
&= \{\ldots, -6, -3, 0, 3, 6, 9, \ldots\}. \\
[1] &= \{x \in Z/x \; R \; 1\} \\
&= \{x \in Z/3 \text{ divides } x - 1\} \\
&= \{3k + 1/k \in Z\} \\
&= \{\ldots, -5, -2, 1, 4, 7, 10, \ldots\}. \\
[2] &= \{x \in Z/x \; R \; 2\} \\
&= \{x \in Z/3 \text{ divides } x - 2\} \\
&= \{3k + 2/k \in Z\} \\
&= \{\ldots, -4, -1, 2, 5, 8, 11, \ldots\}.
\end{aligned}$$

We can see that $[0] = [3] = [6] = [9]$, etc., $[1] = [4] = [7] = [10]$, etc., $[2] = [5] = [8] = [11]$, etc. Thus $[0]$, $[1]$ and $[2]$ are the only distinct equivalence classes. Hence $Z/R = \{[0], [1], [2]\}$, which is denoted by Z_3. Generalising, we obtain $Z_m = \{[0], [1], [2], \ldots, [m-1]\}$.

Example Let $S = \{1, 2, 3, 4, 5\}$ and $A = S \times S$. Define a relation R on A as $(a, b)R(a_1, b_1)$ if and only if $ab_1 = a_1b$. It can be easily seen that R is an equivalence relation. Some of the equivalence classes are

$$[(1,1)] = \{(1,1), (2,2), (3,3), (4,4), (5,5)\}, [(1,2)] = \{(1,2), (2,4)\},$$
$$[(2,1)] = \{(2,1), (4,2)\}, [(1,3)] = \{(1,3)\}, [(3,1)] = \{(3,1)\}, \text{ etc.}$$

Note Let R be an equivalence relation on a finite set A. To compute A/R, repeat the following steps.

Step 1: Take any element $x \in A$ and find $[x]$. If $[x] = A$ then $A/R = \{[x]\}$. Otherwise go to step 2.

Step 2: Take an element $y \in A$ such that $y \notin [x]$. Find $[y]$. If $[x] \cup [y] = A$ then $A/R = \{[x], [y]\}$. Otherwise go to step 3.

Step 3: Take an element $z \in A$ such that $z \notin [x] \cup [y]$. Repeat the above steps until the union of all equivalence classes is A.

6.6 Manipulation of Relations

Consider two sets A and B. Let R and S be two relations from A to B. Then R and S are subsets of $A \times B$. Hence we can form their *union* $R \cup S$. It is the set of all those elements of $A \times B$ which are either in R or in S. Since $R \cup S$ is a subset of $A \times B$; it is also a relation from A to B. This relation is defined as follows.

If $a \in A$ and $b \in B$, then $a(R \cup S)b$ if and only if either aRb or aSb.

We can also form the *intersection* of the two sets R and S. It is the set of all those elements of $A \times B$ which belong to both R and S. Since $R \cap S$ is a subset of $A \times B$, it is also a relation from A to B. This relation is defined as follows.

If $a \in A$ and $b \in B$, then $a\ (R \cap S)\ b$ if and only if aRb and aSb.

Since R is a subset of $A \times B$, we can form its *complement* R^c. It is the set of all those elements of $A \times B$ which do not belong to R. Since R^c is a subset of $A \times B$, it is also a relation from A to B. This relation is defined as follows.

If $a \in A$ and $b \in B$, then aR^cb if and only if a is not related to b through R.

We can also define a relation R^{-1} from B to A as follows.

If $b \in B$ and $a \in A$, then $bR^{-1}a$ if and only if aRb. In other words, $(b, a) \in R^{-1}$ if and only if $(a, b) \in R$.

Example Let $A = \{1, 2, 3, 4\}$ and let

$$R = \{(1, 1), (1, 3), (2, 3), (2, 4), (2, 2), (3, 3), (4, 4), (3, 1)\} \text{ and}$$
$$S = \{(1, 1), (2, 3), (2, 4), (1, 4), (2, 2), (3, 3), (4, 4)\}.$$

Then

$$R \cup S = \{(1, 1), (1, 3), (1, 4), (2, 3), (2, 4), (2, 2), (3, 3)\ (3, 1), (4, 4)\}.$$
$$R \cap S = \{(1, 1), (2, 3), (2, 4), (2, 2), (3, 3), (4, 4)\}.$$
$$R^c = \{(1, 2), (1, 4), (2, 1), (3, 2), (3, 4), (4, 1), (4, 2), (4, 3)\}.$$
$$R^{-1} = \{(1, 1), (3, 1), (3, 2), (4, 2), (2, 2), (3, 3), (4, 4), (1, 3)\}.$$

Result If R and S are both reflexive relations on a set A, then $R \cup S$ and $R \cap S$ are also reflexive relations on A. If R and S are both symmetric relations on A, then $R \cup S$ and $R \cap S$ are also symmetric relations on A.

Proof Suppose R and S are both reflexive. We will prove that $R \cup S$ is also reflexive. Let $a \in A$. Since R is reflexive $(a, a) \in R$ and since S is reflexive $(a, a) \in S$. Hence $(a, a) \in R \cup S$ and this proves that $R \cup S$ is reflexive. Since $(a, a) \in R$ and $(a, a) \in S$, it follows that $(a, a) \in R \cap S$ for all $a \in A$ and hence $R \cap S$ is reflexive.

To prove that $R \cup S$ is symmetric, assume that $(a, b) \in R \cup S$. We have to prove that $(b, a) \in R \cup S$. From $(a, b) \in R \cup S$, we can conclude that either $(a, b) \in R$ or $(a, b) \in S$. If $(a, b) \in R$, then since R is symmetric. It follows that $(b, a) \in R$ and hence $(b, a) \in R \cup S$. We can argue similarly if $(a, b) \in S$.

If $(a, b) \in R \cap S$, then $(a, b) \in R$ and $(a, b) \in S$. Since both R and S are symmetric, $(b, a) \in R$ and $(b, a) \in S$ and hence $(b, a) \in R \cap S$. This proves that $R \cap S$ is symmetric if R and S are both symmetric.

Let R be a relation on a set A and let $a, b \in A$. A *path of length n* from a to b is a sequence $x_0 = a, x_1, x_2, \ldots, x_{n-1}, x_n = b$ such that $aRx_1, x_1Rx_2, \ldots, x_{n-1}R\, b$. Since $a\ R\ x_1$, in the corresponding digraph of R there is a directed edge from a to x_1. Since $x_1\ R\ x_2$, there is a directed edge from x_1 to x_2. Finally, there exists a directed edge from x_{n-1} to b. Combining all these directed edges, we obtain a directed path from a to b.

Definition Let R be a relation on a set A and let $a, b \in A$. We say aR^nb if there exist $x_1, x_2, \ldots, x_{n-1}$ in A such that $aRx_1, x_1Rx_2, \ldots, x_{n-2}\ R\ x_{n-1}$ and $x_{n-1}Rb$.

In particular, when $n = 2$, we obtain $a R^2 b$ if there exists $x_1 \in A$ such that $a\ R\ x_1$ and $x_1\ R\ b$, i.e., if there exists a directed edge from a to x_1 and a directed edge from x_1 to b. When $n = 3$, we obtain $a\ R^3\ b$ if there exist $x_1, x_2 \in A$ such that aRx_1, x_1Rx_2 and x_2Rb.

Example Let $A = \{1, 2, 3, 4, 5\}$ and consider a relation whose digraph is given below.

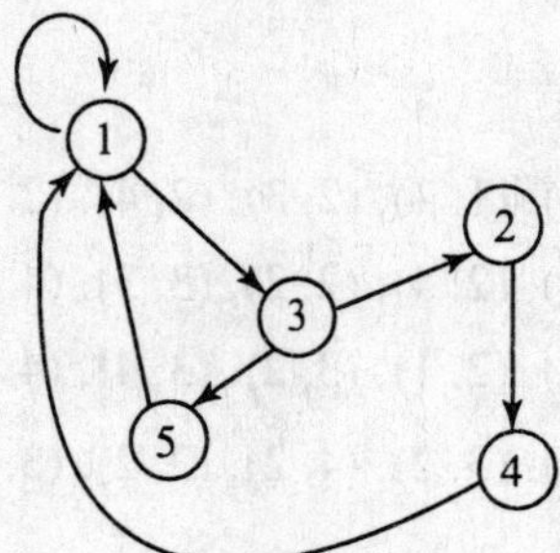

Some of the paths in the above digraph are P_1 : 1, 1, which is a path of length 1 from 1 to 1; P_2 : 1, 3, 2, 4, 1, which is again a path (of length 4) from 1 to 1; and P_3 : 3, 5, 1, which is a path of length 2 from 3 to 1. We note that $1R1$, $1R^4 1$ and $3R^2 1$ but $3R1$ does not exist (no directed edge from 3 to 1).

Paths such as P_1 and P_2 which start from a vertex and end at the same vertex are called *cycles*.

We say that $a R^\infty b$ if there exists a path of arbitrary length from a to b. We have

$$R^n(x) = \{y \in A |\text{ there is a path of length } n \text{ from } x \text{ to } y\}.$$
$$R^\infty(x) = \{y \in A |\text{ there is a path from } x \text{ to } y\}.$$

6.7 Composition of Relations

Consider three sets A, B and C. Let R be a relation from A to B and S be a relation from B to C. We know that R is a subset of $A \times B$ and S is a subset of $B \times C$. We define $S \circ R$ (called composition of R and S) which is a relation from A to C as follows.

If $a \in A$ and $c \in C$, then $a(S \circ R)c$ if and only if there exists $b \in B$ such that aRb and bSc.

Example Let $A = \{1, 2, 3, 4, 5\}$, $B = \{a, b, c, d\}$ and $C = \{x, y, z, u\}$. Define a relation R from A to B and a relation S from B to C as follows.

$$R = \{(1, a), (1, b), (2, b), (2, c), (2, d), (3, b), (3, d), (4, a),$$
$$(4, b), (5, b), (5, c), (5, a)\}.$$
$$S = \{(a, x), (a, y), (a, z), (b, x), (b, z), (c, y), (c, z), (d, x), (d, y)\}.$$

Then,

$$S \circ R = \{(1, x), (1, y), (1, z), (2, x), (2, z), (2, y), (3, x), (3, z), (3, y),$$
$$(4, x), (4, z), (4, y), (5, x), (5, z), (5, y)\}.$$

A matrix all of whose entries are 0s or 1s is called a boolean matrix.

Let $A = (a_{ij})$ be an $m \times n$ boolean matrix and let $B = (b_{ij})$ be an $n \times p$ boolean matrix. We define the boolean product of A and B denoted by $A \otimes B$ to be an $m \times p$ boolean matrix (c_{ij}) where

$$c_{ij} = 1 \text{ if } a_{ik} = 1 \text{ and } b_{kj} = 1 \text{ for some } k (1 \le k \le n).$$
$$= 0 \text{ otherwise.}$$

Result Let R be a relation from A to B and S be a relation from B to C. Then

$$M_{S \circ R} = M_R \circ M_S.$$

Result Let R be a relation on a set A. Let M_R denote the matrix corresponding to relation R and $M_R{}^n$ denote the matrix corresponding to relation R^n. Then $M_R{}^n = M_R \circ M_R \circ \ldots \circ M_R$ (n times) and $M_R{}^\infty = M_R \vee M_R{}^2 \vee M_R{}^3 \vee \ldots = M_R \vee (M_R)^2 \vee (M_R)^3 \vee \ldots$. If A contains n elements, then $M_R{}^\infty = M_R \vee M_R{}^2 \vee M_R{}^3 \vee \ldots \vee M_R{}^n$.

Example For the above relations R and S, we have

$$M_R = \begin{pmatrix} 1 & 1 & 0 & 0 \\ 0 & 1 & 1 & 1 \\ 0 & 1 & 0 & 1 \\ 1 & 1 & 0 & 0 \\ 1 & 1 & 1 & 0 \end{pmatrix}; \quad M_S = \begin{pmatrix} 1 & 1 & 1 & 0 \\ 1 & 0 & 1 & 0 \\ 0 & 1 & 1 & 0 \\ 1 & 1 & 0 & 0 \end{pmatrix};$$

$$M_R \circ M_S = \begin{pmatrix} 1 & 1 & 1 & 0 \\ 1 & 1 & 1 & 0 \\ 1 & 1 & 1 & 0 \\ 1 & 1 & 1 & 0 \\ 1 & 1 & 1 & 0 \end{pmatrix} = M_{S \circ R}.$$

Exercises

1. Let $A = \{1, 2, 3, 4, 5, 6\}$ and let R be a relation on A whose digraph is given below.

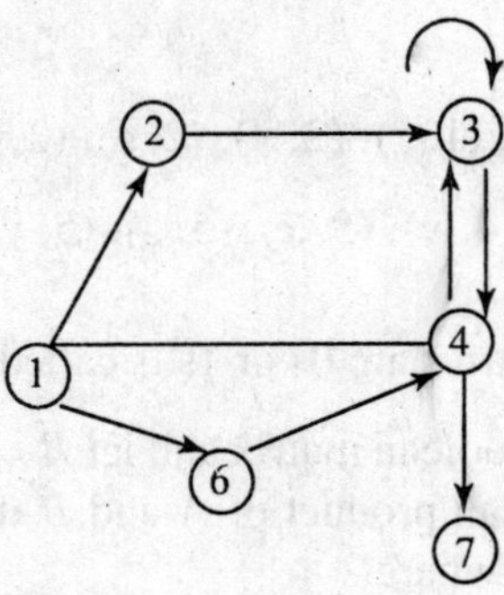

Determine the following.

(a) All paths of length 1.
(b) All paths of length 3 starting from vertex 3.
(c) Cycle starting at vertex 6.
(d) All paths of length 3.
(e) The relation R^∞.

2. Two equivalence relations R and S are given by their relation matrices M_R and M_S. Show that $R \circ S$ is not an equivalence relation.

$$M_R = \begin{bmatrix} 1 & 1 & 0 \\ 1 & 1 & 0 \\ 1 & 0 & 1 \end{bmatrix}; \qquad M_S = \begin{bmatrix} 1 & 1 & 0 \\ 1 & 1 & 1 \\ 0 & 0 & 1 \end{bmatrix}.$$

Obtain equivalence relations R_1 and R_2 on $\{1, 2, 3\}$ such that $R_1 \circ R_2$ is also an equivalence relation.

Consider two boolean matrices A and B (each of size $m \times n$). We define $A \vee B$ to be the $m \times n$ matrix (c_{ij}) where $c_{ij} = a_{ij} \vee b_{ij}$ and the $\vee$ on the right-hand side denotes boolean OR operation. We define $A \wedge B$ to be the $m \times n$ matrix (d_{ij}) where $d_{ij} = a_{ij} \wedge b_{ij}$ and the $\wedge$ on the right-hand side denotes boolean AND operation.

Result Let A and B be two sets and R and S be relations from A to B. Then,

$$M_{R \cup S} = M_R \vee M_S; \; M_{R \cap S} = M_R \wedge M_S.$$

Example Let $A = \{1, 2, 3, 4, 5\}$ and $B = \{a, b, c, d\}$. Consider the following relations from A to B.

$$R = \{(1, a), (1, c), (2, b), (2, c), (3, a), (3, b), (3, d), (4, a), (4, d), (5, b), (5, c)\}.$$

$$S = \{(1, b), (1, c), (2, d), (3, b), (5, a)\}.$$

$$R \cup S = \{(1, a), (1, b), (1, c), (2, b), (2, c), (2, d), (3, a), (3, b), (3, d), (4, a), (4, d), (5, a), (5, b), (5, c)\}.$$

$$R \cap S = \{(1, c), (3, b)\}$$

$$M_R = \begin{bmatrix} 1 & 0 & 1 & 0 \\ 0 & 1 & 1 & 0 \\ 1 & 1 & 0 & 1 \\ 1 & 0 & 0 & 1 \\ 0 & 1 & 1 & 0 \end{bmatrix} \qquad M_S = \begin{bmatrix} 0 & 1 & 1 & 0 \\ 0 & 0 & 0 & 1 \\ 0 & 1 & 0 & 0 \\ 0 & 0 & 0 & 0 \\ 1 & 0 & 0 & 0 \end{bmatrix}$$

$$M_R \wedge M_S = \begin{bmatrix} 0 & 0 & 1 & 0 \\ 0 & 0 & 0 & 0 \\ 0 & 1 & 0 & 0 \\ 0 & 0 & 0 & 0 \\ 0 & 0 & 0 & 0 \end{bmatrix} = M_{R \cap S}$$

$$M_R \vee M_S = \begin{bmatrix} 1 & 1 & 1 & 0 \\ 0 & 1 & 1 & 1 \\ 1 & 1 & 0 & 1 \\ 1 & 0 & 0 & 1 \\ 1 & 1 & 1 & 0 \end{bmatrix} = M_{R \cup S}$$

We define the *reachability relation* R^* on A as follows.

If $a, b \in A$, then aR^*b if and only if either $a = b$ or aR^nb for some n. In other words, b is reachable from a if either $b = a$ or there is a path of length n (for some n) from a to b. We can see that $M_R{}^* = I_n \vee M_R{}^\infty$. Here I_n denotes the $n \times n$ identity matrix.

$$\begin{aligned} M_R{}^* &= I_n \vee M_R{}^\infty \\ &= I_n \vee (M_R \vee M_R{}^2 \vee M_R{}^3 \vee \ldots \vee M_R{}^n). \end{aligned}$$

Definition Let R be a relation on a set A. The *transitive closure* of R is defined to be the smallest transitive relation containing R.

Theorem The transitive closure of R is R^∞.

Proof We will prove the following.

(i) The relation R^∞ is transitive.

(ii) R is contained in R^∞.

(iii) If S is any transitive relation such that $R \subseteq S$, then $R^\infty \subseteq S$.

If we prove the above, then it will follow that R^∞ is the smallest transitive relation containing R.

To prove R^∞ is transitive, assume that $xR^\infty y$ and $yR^\infty z$. Then there is a path in R from x to y, say $x = x_1, x_2, x_3, \ldots, x_{m-1}, x_m = y$. Also, there is a path in R from y to z, say $y = y_1, y_2, y_3, \ldots, y_{n-1}, y_n = z$. Combining these two paths we obtain $x = x_1, x_2, x_3, \ldots, x_{m-1}, y, y_2, \ldots, y_{n-1}, y_n = z$, which is a path from x to z. This proves that $xR^\infty z$ and R^∞ is transitive.

To prove $R \subseteq R^\infty$, assume that $(a, b) \in R$. Then there is a directed edge from a to b which is also a path from a to b. Since a path from a to b exists, it follows that $(a, b) \in R^\infty$ proving that $R \subseteq R^\infty$.

To prove (iii), assume that S is a transitive relation such that $R \subseteq S$. We first note that $R^\infty \subseteq S^\infty$. For, if $(a, b) \in R^\infty$, then there is a path in R from a to b. Since $R \subseteq S$, this path in R from a to b is also a path in S from a to b, proving that $(a, b) \in S^\infty$. We will now prove that $S^\infty \subseteq S$, from which it will follow that $R^\infty \subseteq S$. To prove $S^\infty \subseteq S$, assume that $(a, b) \in S^\infty$. Then there is a path in S from a to b, say $a = a_1, a_2, a_3, \ldots, a_{k-1}, a_k = b$. Then $(a, a_2) \in S, (a_2, a_3) \in S, \ldots, (a_{k-1}, b) \in S$. Since S is transitive, it follows that $(a, b) \in S$. This proves that $S^\infty \subseteq S$.

Solved problem

Let $A = \{1, 2, 3\}$ and $R = \{(1, 1), (1, 2), (2, 3), (1, 3), (3, 1), (3, 2)\}$. Compute the transitive closure of R.

By the above theorem, we have to compute R^∞. Since A contains only three elements, we have $M_{R^\infty} = M_R \vee M_R{}^2 \vee M_R{}^3$. Now,

$$M_R = \begin{pmatrix} 1 & 1 & 1 \\ 0 & 0 & 1 \\ 1 & 1 & 0 \end{pmatrix}.$$

$$M_R{}^2 = M_R \circ M_R = \begin{pmatrix} 1 & 1 & 1 \\ 0 & 0 & 1 \\ 1 & 1 & 0 \end{pmatrix} \circ \begin{pmatrix} 1 & 1 & 1 \\ 0 & 0 & 1 \\ 1 & 1 & 0 \end{pmatrix} = \begin{pmatrix} 1 & 1 & 1 \\ 1 & 1 & 0 \\ 1 & 1 & 1 \end{pmatrix}.$$

$$M_R{}^3 = M_R{}^2 \circ M_R = \begin{pmatrix} 1 & 1 & 1 \\ 1 & 1 & 0 \\ 1 & 1 & 1 \end{pmatrix} \circ \begin{pmatrix} 1 & 1 & 1 \\ 0 & 0 & 1 \\ 1 & 1 & 0 \end{pmatrix} = \begin{pmatrix} 1 & 1 & 1 \\ 1 & 1 & 1 \\ 1 & 1 & 1 \end{pmatrix}$$

It follows that $M_R{}^{\infty} = \begin{pmatrix} 1 & 1 & 1 \\ 1 & 1 & 1 \\ 1 & 1 & 1 \end{pmatrix}$.

Hence $R^{\infty} = \{(1, 1), (1, 2), (1, 3), (2, 1), (2, 2), (2, 3), (3, 1), (3, 2), (3, 3)\}$.

6.8 Warshall's Algorithm

Let A be a set consisting of n elements say $A = \{a_1, a_2, a_3, \ldots, a_n\}$ and let R be a relation on A. Consider a path $x_1, x_2, x_3, \ldots, x_m$ in R. Here x_1 and x_m are said to be the end vertices of the path. All other vertices are called interior vertices. Consider any integer $k(1 \leq k \leq n)$. We will define a boolean matrix W_k as follows.

The ith row, jth column entry of W_k is 1 if and only if there is a path from a_i to a_j in R whose interior vertices come only from the set $\{a_1, a_2, a_3, \ldots, a_k\}$.

We see that the ith row and jth column entry of W_n is 1 if and only if there is a path from a_i to a_j in R whose interior vertices come from the set $A = \{a_1, a_2, \ldots, a_n\}$. Since all interior vertices have to come only from the set $\{a_1, a_2, \ldots, a_n\}$, it follows that the ith row, jth column entry of W_n is 1 if and only if there is a path from a_i to a_j in R if and only if the ith row and jth column entry of $M_R{}^{\infty}$ is 1. This means $W_n = M_R{}^{\infty}$.

We take $W_0 = M_R$ and show how to obtain W_k if W_{k-1} is known. This will enable us to obtain $W_1, W_2, \ldots, W_{n-1}, W_n$. Finally, W_n will give the transitive closure of R.

Let $W_{k-1} = (q_{ij})$ and $W_k = (p_{ij})$. Suppose $p_{ij} = 1$. Then by the definition of W_k there is a path from a_i to a_j whose interior vertices come from the set $\{a_1, a_2, \ldots, a_k\}$. Suppose a_k is not an interior vertex of the path. Then this path from a_i to a_j has its interior vertices coming from the set $\{a_1, a_2, \ldots, a_{k-1}\}$, proving that $q_{ij} = 1$. Conversely, we can show that if $q_{ij} = 1$, then $p_{ij} = 1$.

Suppose a_k is an interior vertex. Then the path from a_i to a_j passes through a_k and can be split into two parts: one from a_i to a_k and the other from a_k to a_j. The path from a_i to a_k has all its interior vertices coming from the set $\{a_1, a_2, \ldots, a_{k-1}\}$ and the path from a_k to a_j also has all its interior vertices coming from the same set. Hence $q_{ik} = 1$ and $q_{kj} = 1$. Conversely, if $q_{ik} = 1$ and $q_{kj} = 1$, then $p_{ij} = 1$. In other words, to obtain the ith row and jth column entry of W_k, we look at the ith row and the jth column entry of W_{k-1}. If the latter entry is 1, then 1 is placed in the ith row, jth column entry of W_k. If it is not 1, then we look at the ith row, kth column entry and kth row, jth column entry of W_{k-1}. If both are 1, we place 1 in the ith row and jth column position of W_k. All those entries of W_k where there are no 1s are made 0s.

Example Consider the set A and the relation R given in the above problem. Compute the transitive closure of R using Warshall's algorithm.

Solution We have

$$W_0 = M_R = \begin{bmatrix} 1 & 1 & 1 \\ 0 & 0 & 1 \\ 1 & 1 & 0 \end{bmatrix}.$$

Applying Warshall's algorithm, we obtain

$$W_1 = \begin{bmatrix} 1 & 1 & 1 \\ 0 & 0 & 1 \\ 1 & 1 & 1 \end{bmatrix}; \quad W_2 = \begin{bmatrix} 1 & 1 & 1 \\ 0 & 0 & 1 \\ 1 & 1 & 1 \end{bmatrix};$$

$$W_3 = \begin{bmatrix} 1 & 1 & 1 \\ 1 & 1 & 1 \\ 1 & 1 & 1 \end{bmatrix} = M_{R^\infty}.$$

Consider a set A and two equivalence relations R and S on A. Then clearly $R \cap S$ is an equivalence relation on A. We also know $R \cap S$ is contained in R and $R \cap S$ is contained in S. In fact, $R \cap S$ is the largest equivalence relation which is contained in both R and S. This follows from the fact that if T is an equivalence relation such that T is contained in R and T is contained in S, then T is contained in $R \cap S$.

Now consider $R \cup S$. Since R and S are both reflexive and symmetric, $R \cup S$ is also reflexive and symmetric. For all $a \in A$, $(a, a) \in R$ and $(a, a) \in S$ since R and S are both reflexive; hence $(a, a) \in R \cup S$ proving that $R \cup S$ is reflexive. If $(a, b) \in R \cup S$, then $(a, b) \in R$ or $(a, b) \in S$. If $(a, b) \in R$, then $(b, a) \in R$ because R is symmetric. If $(a, b) \in S$, then $(b, a) \in S$ because S is also symmetric. In either of these two cases, $(b, a) \in R \cup S$. This proves that $R \cup S$ is symmetric.

But $R \cup S$ need not be transitive. If $(a, b) \in R \cup S$ and $(b, c) \in R \cup S$, then it may happen that $(a, b) \in R$ and $(b, c) \in S$ or $(a, b) \in S$ and $(b, c) \in R$. From this we cannot conclude that $(a, c) \in R \cup S$. Hence $R \cup S$ need not be an equivalence relation.

Now consider $(R \cup S)^\infty$ which is the smallest transitive relation containing $R \cup S$. We can see that $(R \cup S)^\infty$ is the smallest relation containing both R and S.

7 Functions

7.1 Introduction

In this chapter we consider the definitions of a function, one–one function and an onto function. We also consider the concept of pigeon hole principle and the various applications that this innocent looking concept has.

One–one and onto functions have several interesting applications. Suppose we want to count the number of elements in a finite set A. It may be very difficult to do so. In such cases, we can find another finite set B where it is easy to count the number of elements. We then establish a function from A to B which is both one–one and onto. Then the number of elements in A will be equal to the number of elements in B. Such examples are abundant in combinatorics. We will see one such example later in this chapter.

Definition Let A and B be two sets. A function from A to B is a rule which assigns to every element $a \in A$ one and only one element of B. If f is a function from A to B, then the element of B which is associated with an element $a \in A$ (we know this element in B exists and is unique) is denoted by $f(a)$ and is called the image of a. If $b = f(a)$, then a is called a pre-image of b. This function is denoted by writing $f\colon A \to B$. A is called the domain of the function and B is called the co-domain of the function. We define the range of f denoted by $\text{Ran}(f)$ to be the set of all images. In other words, $\text{Ran}(f) = \{f(a) | a \in A\}$. $\text{Ran}(f)$ is a subset of the co-domain.

Note

1. In the definition of a function given by some people it is not necessary that an element of B must be associated with every element of A. In other words, some elements of A may not be associated with any element of B. If however every element of A is associated with an element of B, such functions are called *everywhere defined* functions (functions which are defined for every element of A). We will assume that all our functions

are everywhere defined. Thus in our definition of function an element of B is associated with every element of A.

2. More than one element of B cannot be associated with an element $a \in A$. This means that every element of A has a unique image in B.
3. It is possible that different elements of A are associated with the same element of B.
4. It is not necessary that with every element $b \in B$ there is an element $a \in A$ which is associated with b. In other words, some elements of B may be left out.
5. Pre-image of an element need not be unique. Given an element $b \in B$, there can exist zero, one or more than one element of A which are associated with b.

Definition If $f: A \to B$ is a function, then for an element $b \in B$ we define $f^{-1}(b) = \{a \in A | f(a) = b\}$.

From Note 5 it is observed that $f^{-1}(b)$ is a subset of A containing zero, one or more than one element.

Examples

1. Let A = {a, b, c, d} and B = {1, 2, 3}.
   Define $f: A \to B$ as $f(a) = 2$, $f(b) = 3$, $f(c) = 1$ and $f(d) = 2$.
   Then f is a function from A to B. Note that two distinct elements a and d in A are associated with one element 2 in B. Note also that no element of B is left out. Corresponding to 1 in B there exists an element c in A such that $f(c) = 1$. Corresponding to 2 in B there exist two elements a and d in A which are associated with 2. Corresponding to 3 in B there is an element b in A which is associated with 3. $\text{Ran}(f) = \{1, 2, 3\} = B$. We have $f^{-1}(2) = \{a, d\}$, $f^{-1}(3) = \{b\}$ and $f^{-1}(1) = \{c\}$.
   The above function can also be denoted as $\{(a, 2), (b, 3), (c, 1), (d, 2)\}$ (the set of all ordered pairs the first component of each being an element of A and the second component its image under f), which is nothing but a subset of $A \times B$ and hence a relation from A to B. Thus every function from A to B is also a relation from A to B.
2. Consider the sets A and B given above and the relation from A to B given by {(a, 3), (b, 1), (b, 2), (c, 1), (d, 3)}. This relation is not a function from A to B since the element $b \in A$ is associated with two distinct elements 1 and 2 in B, which cannot happen in a function.
   From the above two examples we see that a function f from A to B is also a relation from A to B, whereas a relation from A to B need not be a function from A to B.

3. Let $X = \{\text{cat, dog, tiger}\}$, $Y = \{\text{rice, wheat, dal, ragi}\}$. g is defined as $g(\text{cat}) = \text{wheat}$, $g(\text{dog}) = \text{ragi}$ and $g(\text{tiger}) = \text{dal}$. Then g is a function from X to Y.

   In the above example we note that distinct elements of X are associated with distinct elements of Y but some elements of Y are left out. For example, consider rice $\in Y$. There does not exist any element $x \in X$ such that $g(x) = \text{rice}$. We have $\text{Ran}(g) = \{\text{wheat, dal, ragi}\} \subset Y$. Also, $g^{-1}(\text{wheat}) = \{\text{cat}\}$, $g^{-1}(\text{ragi}) = \{\text{dog}\}$, $g^{-1}(\text{dal}) = \{\text{tiger}\}$ and $g^{-1}(\text{rice}) = \phi$.

   The above function can be pictorially represented as follows:

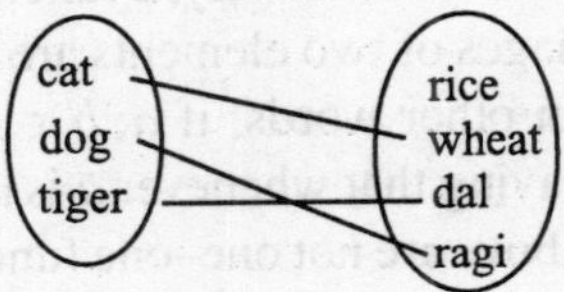

4. Let $A = \{a, b, c, d\}$. Consider the function $h: A \to A$ defined as $h(a) = a$, $h(b) = b$, $h(c) = c$, $h(d) = d$.

   The pictorial representation of this function is given below.

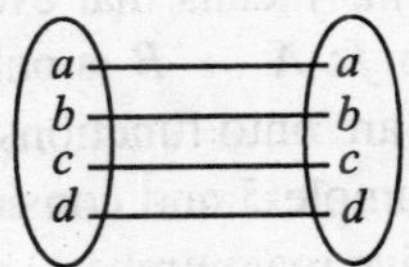

   This function associates (we also say maps) each element of A with itself. It is called identity function on A and is denoted by I_A. Note that for this function the domain and the co-domain are A itself. Any two distinct elements of A are associated with distinct elements of A and corresponding to any element $x \in A$ there exists an element in A (x itself) which is associated with x. We have $\text{Ran}(h) = A$, $h^{-1}(a) = \{a\}$, $h^{-1}(b) = \{b\}$, $h^{-1}(c) = \{c\}$ and $h^{-1}(d) = \{d\}$.

5. Let $\mathbb{R}$ denote the set of all real numbers and consider the function f from $\mathbb{R}$ to $\mathbb{R}$ defined as $f(x) = x^2$. This function associates with every real number a real number which is its square. For example, $f(-8) = 64 = f(8)$, $f(11) = 121 = f(-11)$ and so on. Note that there are distinct real numbers (for example, 15 and -15) which are associated with the same real number 225. If we take any positive real number, then there is another real number (its square root) which is associated with the given positive real number. For example, given the positive real number 21, there are two real numbers $\sqrt{21}$ and $-\sqrt{21}$ which are associated with 21. $f(\sqrt{21}) = f(-\sqrt{21}) = 21$. However, for negative real numbers

there are no corresponding real numbers which are associated with those negative real numbers. This is because the square root of negative real numbers are not real numbers.

6. Let $A = \{a, b, c\}$ and $B = \{1, 2, 3\}$. Consider the function $f\colon A \to B$ defined as $f(a) = 2$, $f(b) = 2$ and $f(c) = 2$. Then f is a function called constant function.

We have already seen that in a function f from A to B, distinct elements of A may be associated with the same element of B (Examples 1, 5 and 6). If distinct elements of A are associated with distinct elements of B, then the function is said to be one–one or injective. Alternatively, a function from A to B is said to be one–one if whenever the images of two elements are the same, then the elements themselves are the same. In other words, if $a, b \in A$ and $f(a) = f(b)$, then $a = b$. This is the same as saying that whenever a is not equal to b, $f(a) \neq f(b)$. Examples 1, 5 and 6 given above are not one–one functions, whereas Examples 3 and 4 are.

We have also seen that in a function $f\colon A \to B$, given an element $b \in B$, there need not exist any element $a \in A$ such that $f(a) = b$. A function $f\colon A \to B$ is said to be onto (surjective) if for every element $b \in B$ there exists an element $a \in A$ such that $f(a) = b$. This means that every element $b \in B$ has a pre-image in A. Clearly, a function $f\colon A \to B$ is onto if and only if $\text{Ran}(f) = B$. Examples 1 and 4 given above are onto functions whereas Examples 3, 5 and 6 are not onto. If we modify Example 5 and consider the function $f_1\colon \mathbb{R} \to \mathbb{R}^+$ ($\mathbb{R}^+$ denotes the set of all positive real numbers) defined as $f_1(x) = x^2$, then f_1 is onto. This is because every positive real number has two square roots which are real numbers and are hence associated with the given positive real number. However this function is not one–one.

From the above examples we note that there are functions which are (i) onto but not one–one (Example 1), (ii) one–one but not onto (Example 3), (iii) both one–one and onto (Example 4) and (iv) neither one–one nor onto (Examples 5 and 6). However the following is true.

Result Let A and B be two finite sets consisting of the same number of elements. Then a function f from A to B is one–one if and only if it is onto.

Proof Let $|A| = |B| = n$. Suppose $f\colon A \to B$ is one–one. Then by the definition of one–one function, n elements of A must be associated with minimum n elements of B (since no two elements of A can be associated with the same element of B). Also, no element of A can be associated with more than one element of B (by definition of a function). Hence it follows that n elements of A are associated with n elements of B. Since B contains only n elements, no element of B is left out, proving that f: $A \to B$ is onto. Similarly assuming that f: $A \to B$ is onto, we can prove that f is one–one.

This result fails for Examples 1 and 3 given above since the number of elements in domain and co-domain are not the same.

Note

1. If the number of elements in A is more than the number of elements in B (as in Example 1), then there cannot exist a function from A to B which is one–one. This is obvious because if distinct elements of A are to be associated with distinct elements of B, then B must have at least as many elements as A has.
2. If the number of elements in X is less than the number of elements in Y (as in Example 3), then there cannot exist a function from X to Y which is onto. If $|X| = m$, $|Y| = n$ and $m < n$, then for m specified elements in Y we can find m pre-images in X (the number of pre-images cannot be less than m since an element of X cannot have two images in Y by definition of a function). The remaining $n - m$ (greater than zero since $m < n$) elements of Y do not have pre-images in X. If at all they have pre-images, they must be among the m pre-images we have already found (since X has only m elements). In this case one element of X will have two distinct images in Y which is against the definition of a function.

From Note 1 it follows that if a function f: $A \to B$ is one–one, then the number of elements in A must be less than or equal to the number of elements in B. From Note 2 it follows that if a function f: $A \to B$ is onto, then the number of elements in A must be greater than or equal to the number of elements in B. Combining the two, we obtain that if a function f: $A \to B$ is both one–one and onto, then the number of elements in A must be equal to the number of elements in B. Conversely, if A and B contain the same number of elements, then we can define a function f: $A \to B$ which is one–one (and hence onto also by the above result). We can also define a function g: $A \to B$ which is onto (and hence one–one by the above result).

Definition A function f: $A \to B$ which is both one–one and onto is called one–one correspondence or a bijection. A function p: $A \to A$ which is both one–one and onto is called a permutation on A.

Examples

1. Example 4 given above is a permutation on A called the identity permutation on A.

2. Let $A = \{a, b, c\}$ and consider a function $p: A \to A$ defined as $p(a) = c$, $p(b) = a$ and $p(c) = b$. Then p is a permutation on A. Its pictorial representation is given below.

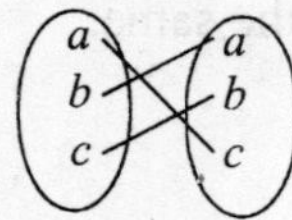

Note

1. Whenever we want to prove that a function $f: A \to B$ is one–one, we have to prove that if $f(a) = f(b) (a, b \in A)$, then $a = b$.
2. To prove that f is not one–one, we have to find elements $a, b \in A$, $a \neq b$ but $f(a) = f(b)$.
3. To prove that a function $f: A \to B$ is onto, we have to prove that for every $b \in B$, there exists $a \in A$ such that $f(a) = b$.
4. To prove that f is not onto, we have to find an element $b \in B$ for which there does not exist any element $a \in A$ such that $f(a) = b$.

Let $f: A \to B$ and $g: B \to C$ be two functions. We define a function $g \circ f$: $A \to C$ as follows.

Let $a \in A$. Then $f(a) \in B$ and hence $g(f(a)) \in C$. Define $g \circ f(a) = g(f(a))$. $g \circ f$ is called the *composition* of f and g.

Note $g \circ f$ is defined only when the co-domain of f is the same as the domain of g.

Two functions f and g are said to be equal ($f = g$) if and only if the domain of f is equal to the domain of g, the co-domain of f is equal to the co-domain of g and $f(x) = g(x)$ for all x.

If $f: A \to B$, $g: B \to C$ and $h: C \to D$ are functions, then $h \circ (g \circ f)$ is defined and is a function from A to D. $(h \circ g) \circ f$ is also defined and it is a function from A to D. For every $a \in A$, we have

$$(h \circ (g \circ f))(a) = h(g \circ f(a)) = h(g(f(a))).$$

Again, $((h \circ g) \circ f)(a) = (h \circ g)(f(a)) = h(g(f(a))) = (h \circ (g \circ f))(a)$. From the above, $h \circ (g \circ f) = (h \circ g) \circ f$, proving that $\circ$ is associative.

If $g \circ f$ exists, then $f \circ g$ may or may not exist. ($f \circ g$ will exist only if the co-domain of g is the same as the domain of f). Even if both $f \circ g$ and $g \circ f$ exist, it is not necessary that $f \circ g = g \circ f$. Hence $\circ$ is not commutative.

Definition Let $f: A \to B$ be a function. A function $g: B \to A$ is said to be an *inverse* of f if $g \circ f = I_A$ and $f \circ g = I_B$. That is, for all $a \in A$, $(g \circ f)(a) = a$ and for all $b \in B$, $(f \circ g)(b) = b$.

Note It is not necessary that a function $f: A \to B$ must have an inverse. If, however, f has an inverse, then its inverse is unique. This unique inverse of f is denoted by f^{-1}. In this case we say that f is *invertible*.

Result A function $f: A \to B$ is invertible if and only if f is one–one and onto.

Proof First assume that $f: A \to B$ is invertible. Then, by definition there exists a function $g: B \to A$ such that $g \circ f = I_A$ and $f \circ g = I_B$. If $f(x) = f(y)$ where $x, y \in A$, then $g(f(x)) = g(f(y))$. (When two elements are the same, their images under a function must be the same.) By definition of composition of functions, $g \circ f(x) = g \circ f(y)$. This means $I_A(x) = I_A(y)$ which is the same as saying that $x = y$. Hence f is one–one. To prove that $f: A \to B$ is onto, consider any element $b \in B$. Now $b = I_B(b) = f \circ g(b) = f(g(b))$. We have thus found an element $a = g(b) \in A$ such that $b = f(a)$, proving that f is onto.

Conversely, assume that $f: A \to B$ is both one–one and onto. We will find a function $g: B \to A$ such that $g \circ f = I_A$ and $f \circ g = I_B$. Let $b \in B$. Since f: $A \to B$ is onto, there exists $a \in A$ such that $f(a) = b$. Since $f: A \to B$ is one–one, this element a is unique. Define $g(b) = a$. Let $x \in A$ and let $f(x) = y \in B$. By definition of g, we have $g(y) = x$. Now $g \circ f(x) = g(f(x)) = g(y) = x$ for all $x \in A$, proving that $g \circ f = I_A$. Similarly, we can prove that $f \circ g = I_B$. This completes the proof.

Solved problems

1. Let $f: \mathbb{R} \to \mathbb{R}$ and $g: \mathbb{R} \to \mathbb{R}$ be defined as $f(a) = a - 1$ and $g(b) = b^2$ Determine

   (i) $(f \circ g)$ (2)
   (ii) $(g \circ f)(2)$

   Solution

   (i) $(f \circ g)(2) = f(g(2)) = f(2^2) = f(4) = 4 - 1 = 3$.
   (ii) $(g \circ f)(2) = g(f(2)) = g(2 - 1) = g(1) = 1^2 = 1$.

   We note that both $f \circ g$ and $g \circ f$ exist but $f \circ g \neq g \circ f$.

2. If $f: A \to B$ and $g: B \to C$ are one–one functions, prove that $g \circ f$ is one–one.

Solution We know that $g \circ f$ is a function from A to C. To prove $g \circ f$ is one–one, assume that $g \circ f(x) = g \circ f(y)$, where $x, y \in A$. By definition of composition of functions, we have $g(f(x)) = g(f(y))$. Since g is one–one, we have $f(x) = f(y)$. Again, since f is one–one, it follows that $x = y$ proving that $g \circ f$ is one–one.

3. If $f: A \to B$ and $g: B \to C$ are onto functions, prove that $g \circ f$ is onto.

   Solution To prove $g \circ f: A \to C$ is onto, consider $c \in C$. We have to find $a \in A$ such that $g \circ f(a) = c$. Since $g: B \to C$ is onto, given $c \in C$, we can find $b \in B$ such that $g(b) = c$. Since $f: A \to B$ is onto, given $b \in B$, we can find $a \in A$ such that $f(a) = b$. Now, $g \circ f(a) = g(f(a)) = g(b) = c$ proving that $g \circ f: A \to C$ is onto.

4. Let $f: A \to B$ and $g: B \to C$ be two functions. If $g \circ f: A \to C$ is one–one, prove that f is one–one.

   Solution To prove that $f: A \to B$ is one–one, assume that

   $$f(x) = f(y)(x, y \in A).$$

   We have to prove that $x = y$. Since $f(x) = f(y)$, by definition of a function it follows that $g(f(x)) = g(f(y))$, i.e., $g \circ f(x) = g \circ f(y)$. Since $g \circ f$ is one–one, we have $x = y$ proving that f is one–one.

5. Let $f: A \to B$ and $g: B \to C$ be two functions. If $g \circ f: A \to C$ is onto, prove that g is onto.

   Solution Let $c \in C$. We have to find $b \in B$ such that $g(b) = c$. Since $g \circ f: A \to C$ is onto, given $c \in C$, we can find $a \in A$ such that $g \circ f(a) = c$. This means $g(f(a)) = c$. Let $f(a) = b \in B$. Then $g(b) = c$ proving that $g: B \to C$ is onto.

Exercises

1. In each of the following problems determine whether the function $f: A \to B$ is one–one and whether it is onto.

   (i) $A = B = Z$, $f(a) = a - 1$

   (ii) $A = B = \mathbb{R}$, $f(a) = |a|$

   (iii) $A = \mathbb{R} \times \mathbb{R}$, $B = \mathbb{R}$, $f((a, b)) = a$

   (iv) $A = B = \mathbb{R} \times \mathbb{R}$, $f((a, b)) = (a + b, a - b)$

2. Let N denote the set of all natural numbers including zero. Determine which of the following functions are one–one, which are onto and which are one–one and onto.

(a) $f: N \to N$ $\quad f(j) = j^2 + 2$

(b) $f: N \to N$ $\quad f(j) = j \pmod 3$

(c) $f: N \to N$ $\quad f(j) = 1$ when j is odd, 0 when j is even.

(d) $f: N \to \{0, 1\}$ $\quad f(j) = 1$ when j is odd, 0 when j is even.

3. Let Z denote the set of all integers, Z^+ the set of all positive integers and $Z_p = \{0, 1, 2, 3, \ldots, p-1\}$. Determine which of the following functions are one–one, which are onto and which are one–one and onto.

   (a) $f: Z \to Z$ defined as $f(j) = \begin{cases} \frac{j}{2} & \text{when } j \text{ is even,} \\ \frac{(j-1)}{2} & \text{when } j \text{ is odd.} \end{cases}$

   (b) $f: Z^+ \to Z^+$ defined as $f(x) =$ greatest integer $\leq \sqrt{x}$.

   (c) $f: Z_7 \to Z_7$ defined as $f(x) = 3x \pmod 7$, the remainder obtained when $3x$ is divided by 7.

   (d) $f: Z_4 \to Z_4$ defined as $f(x) = 3x \pmod 4$.

4. If X and Y are finite sets, find a necessary condition for the existence of one–one mapping from X to Y.
5. Let $A = \{a, b, c, d\}$ and $B = \{1, 2, 3\}$. Determine whether each of the relations R from A to B given below is a function. If it is, give its range.

   (i) $R = \{(a, 3), (b, 2), (a, 1), (c, 1)\}$

   (ii) $R = \{(a, 1), (b, 1), (c, 1), (d, 1)\}$

6. List all possible functions from $X = \{a, b, c\}$ to $Y = \{0, 1\}$ and indicate in each case whether the function is one–one, onto or, one–one and onto.
7. Show that the functions f and g from $N \times N$ to N given by $f(x, y) = x + y$ and $g(x, y) = xy$ (xy denotes the product of x and y) are onto but not one–one.
8. Determine the number of functions from X to Y where X and Y are the sets given below. Find also the number of functions which are one–one, which are onto and which are one–one and onto.

   (a) $X = \{1, 2, 3\}$ $\quad Y = \{1, 2, 3\}$

   (b) $X = \{1, 2, 3, 4\}$ $\quad Y = \{1, 2, 3\}$

   (c) $X = \{1, 2, 3\}$ $\quad Y = \{1, 2, 3, 4\}$

9. In each of the following problems, determine whether f is invertible. If so, determine f^{-1}.

(i) $f: \mathbb{R} \rightarrow \mathbb{R}$ defined as $f(a) = a^3 + 1$

(ii) $g: \mathbb{R} \rightarrow \mathbb{R}$ defined as $g(a) = \frac{(2a-1)}{3}$

(iii) $f: \mathbb{R} \rightarrow \mathbb{R}$ defined as $f(x) = x^2 - 2$

10. Let $f: \mathbb{R} \rightarrow \mathbb{R}$ and $g: \mathbb{R} \rightarrow \mathbb{R}$ be defined as $f(x) = x^2 - 2$ and $g(x) = x + 4$. Find $f \circ g$ and $g \circ f$. State whether these functions are injective, surjective and bijective.

11. Let f, g and h be functions from N to N defined as

$$f(n) = n + 1, \; g(n) = 2n \quad \text{and} \quad h(n) = \begin{cases} 0 \text{ when } n \text{ is even,} \\ 1 \text{ when } n \text{ is odd.} \end{cases}$$

Determine $f \circ f$, $f \circ g$, $g \circ f$, $g \circ h$, $h \circ g$, $(f \circ g) \circ h$.

7.2 Pigeonhole Principle

Assume that m persons $P_1, P_2, P_3, \ldots\ldots, P_m$ go to a hotel for an overnight stay. Suppose the hotel manager informs them that $n(\geq 1)$ rooms $R_1, R_2, R_3, \ldots\ldots, R_n$ are available. We can think of a function f from the set $P = \{P_1, P_2, P_3, \ldots, P_m\}$ to the set $R = \{R_1, R_2, R_3, \ldots, R_n\}$ defined as $f(P_i) = R_j$ if the person P_i is assigned the room R_i. Since at least one room is available, every person will be assigned a room. Hence f is defined for all elements of P. Also, one person will not be assigned two rooms. Hence f is a well-defined function. If the function f is one–one , then whenever $f(P_i) = f(P_k)$, then $P_i = P_k$. That is, if the rooms assigned to two persons are the same, then the persons themselves are the same. This is the same as saying that each person is assigned a different room. Given two different persons, the rooms assigned to these two persons are also different. We know this can happen only when the number of rooms available is greater than or equal to the number of persons. In other words, if the number of persons is more than the number of rooms available, then it is not possible to assign different rooms to different persons. This means that at least two persons will have to share a room. Depending on the number of rooms available (the value of n), two or more persons will have to share rooms. If only one room is available, then all m persons will have to share that one room. This simple result is known as Pigeonhole Principle.

Statement of pigeonhole principle

If n pigeons are assigned to m pigeonholes where $n > m$, then at least one pigeonhole will have to be occupied by more than one pigeon.

Immediate applications of pigeonhole principle

1. In a gathering of eight persons, at least two would have been born on the same day of the week. This is because there are 8 persons but there are only 7 days in a week.
2. In a gathering of 13 persons, at least two would have been born on the same month of the year. This is because there are 13 persons but there are only 12 months in a year.

Notation For any two positive integers m and n, $\left[\frac{m}{n}\right]$ denotes the largest integer not larger than the number $\frac{m}{n}$. Clearly, $\left[\frac{m}{n}\right] \le \frac{m}{n}$. For example. $\left[\frac{37}{4}\right] = 9$.

7.3 Extended Pigeonhole Principle

Extended pigeonhole principle says that if n pigeons are assigned to m pigeonholes, then one of the pigeonholes must have at least $\left[\frac{(n-1)}{m}\right] + 1$ pigeons.

Proof Suppose every pigeonhole contains a maximum of $\left[\frac{(n-1)}{m}\right]$ pigeons. Then m pigeonholes will contain a maximum of $m\left[\frac{(n-1)}{m}\right]$ pigeons. But $m\left[\frac{(n-1)}{m}\right] \le m(\frac{(n-1)}{m}) = n - 1 < n$. Since there are n pigeons, our assumption that every hole contains a maximum of $\left[\frac{(n-1)}{m}\right]$ pigeons is wrong and hence at least one hole must contain a minimum of $\left[\frac{(n-1)}{m}\right] + 1$ pigeons.

Solved problem Show that if 7 numbers are chosen from 1 to 12, then two of them will add to 13.

Solution Form sets each containing two numbers from 1 to 12 such that their sum is 13. The only possible sets are $A_1 = \{1, 12\}$, $A_2 = \{2, 11\}$, $A_3 = \{3, 10\}$, $A_4 = \{4, 9\}$, $A_5 = \{5, 8\}$ and $A_6 = \{6, 7\}$ Note that the set $\{4, 9\}$ is the same as the set $\{9, 4\}$. We see that $A_1 \cup A_2 \cup A_3 \cup A_4 \cup A_5 \cup A_6 = \{1, 2, 3, \ldots, 12\}$. Consider seven numbers $n_1, n_2, n_3, n_4, n_5, n_6, n_7$ chosen from the set $1, 2, 3, \ldots, 12$. Now $n_1 \in \{1, 2, 3, \ldots, 12\} = A_1 \cup A_2 \cup A_3 \cup A_4 \cup A_5 \cup A_6$. Hence n_1 belongs to one of the sets $A_1, A_2, A_3, A_4, A_5, A_6$. Similarly n_2 belongs to one of the sets $A_1, A_2, A_3, A_4, A_5, A_6$. In general, each number $n_i (1 \le i \le 7)$ belongs to one of the sets $A_1, A_2, A_3, A_4, A_5, A_6$. Since there are 7 numbers but only 6 sets $A_1, A_2, A_3, A_4, A_5, A_6$, by pigeonhole principle two of these numbers (say n_k and n_l) must belong to the same set. Then $n_k + n_l = 13$.

Exercises

1. Show that if 13 people are selected at random, there must be five people who are born on the same day of the week.
2. Show that if 8 positive integers are chosen, two of them will have the same remainder when divided by 7.
3. Show that if 7 colours are used to paint 50 cars, at least 8 cars will have the same colour.

7.4 Permutations

We have already seen that a permutation on a set A is nothing but a function from A to itself which is both one–one and onto. If $A = \{a_1, a_2, .., a_n\}$ and p is a permutation on A, then p can be represented as shown below.

$$\begin{pmatrix} a_1 & a_2 & a_3 & \ldots\ldots & a_n \\ p(a_1) & p(a_2) & p(a_3) & \ldots\ldots & p(a_n) \end{pmatrix}$$

Here $p(a_j)$ denotes the image of a_j under $p (j = 1, 2, \ldots, n)$.

Example Let A = {1, 2, 3, 4} and consider a permutation p on A defined as $p(1) = 4$, $p(2) = 2$, $p(3) = 1$ and $p(4) = 3$. Then p is a permutation on A which can be represented as

$$p = \begin{pmatrix} 1 & 2 & 3 & 4 \\ 4 & 2 & 1 & 3 \end{pmatrix}.$$

Consider another permutation q on A given by the following.

$$q = \begin{pmatrix} 1 & 2 & 3 & 4 \\ 3 & 1 & 4 & 2 \end{pmatrix}.$$

Here $q(1) = 3$, $q(2) = 1$, $q(3) = 4$ and $q(4) = 2$.

$$p \circ q = \begin{pmatrix} 1 & 2 & 3 & 4 \\ 1 & 4 & 3 & 2 \end{pmatrix}$$

$$q \circ p = \begin{pmatrix} 1 & 2 & 3 & 4 \\ 2 & 1 & 3 & 4 \end{pmatrix}$$

We note that since both p and q are functions from A to itself, both $p \circ q$ and $q \circ p$ exist, but $p \circ q \neq q \circ p$.

Let $A = \{a_1, a_2, \ldots, a_n\}$ and let p be a permutation on A. Let $a_{i1}, a_{i2}, \ldots, a_{ik} \in A$ such that $p(a_{i1}) = a_{i2}$, $p(a_{i2}) = a_{i3}, \ldots\ldots, p(a_{ik-1}) = a_{ik}$ and $p(a_{ik}) = a_{i1}$ and let $p(a_j) = a_j$ for all other elements of A. The permutation p can be represented as $(a_{i1}, a_{i2}, a_{i3}, \ldots, a_{ik-1}, a_{ik})$ and is called a *cycle* of length k. A cycle of length 2 is called a transposition. Thus a *transposition* is of the form (a_{i1}, a_{i2}). Here $p(a_{i1}) = a_{i2}$ and $p(a_{i2}) = a_{i1}$. $p(a_j) = a_j$ for all other elements.

Note Let $A = \{a_1, a_2, \ldots, a_n\}$ and $p = (a_i, a_j)$ be a transposition. Then,

$$\begin{aligned} p \circ p &= \begin{pmatrix} a_1 & a_2 & a_3 & \ldots & a_{i-1} & a_i & \ldots & a_{j-1} & a_j & \ldots & a_n \\ a_1 & a_2 & a_3 & \ldots & a_{i-1} & a_j & \ldots & a_{j-1} & a_i & \ldots & a_n \end{pmatrix} \\ &\quad \circ \begin{pmatrix} a_1 & a_2 & a_3 & \ldots & a_{i-1} & a_i & \ldots & a_{j-1} & a_j & \ldots & a_n \\ a_1 & a_2 & a_3 & \ldots & a_{i-1} & a_j & \ldots & a_{j-1} & a_i & \ldots & a_n \end{pmatrix} \\ &= \begin{pmatrix} a_1 & a_2 & a_3 & \ldots & a_{i-1} & a_i & \ldots & a_{j-1} & a_j & \ldots & a_n \\ a_1 & a_2 & a_3 & \ldots & a_{i-1} & a_i & \ldots & a_{j-1} & a_j & \ldots & a_n \end{pmatrix} \\ &= I_A. \end{aligned}$$

Example The permutation p given in the above example can be represented by the cycle (1, 4, 3) and the permutation q by the cycle (1, 3, 4, 2).

Two cycles $(a_{i1}, a_{i2}, a_{i3}, \ldots, a_{ik})$ and $(b_{j1}, b_{j2}, b_{j3}, \ldots, b_{jl})$ are said to be *disjoint cycles* if $\{a_{i1}, a_{i2}, a_{i3}, \ldots, a_{ik}\} \cap \{b_{j1}, b_{j2}, b_{j3}, \ldots\ldots, b_{jl}\} = \phi$. That is, no element is common between the cycles. Any permutation can be expressed as a product of disjoint cycles as the following example shows.

Let $A = \{1, 2, 3, 4, 5, 6, 7, 8\}$ and consider a permutation p on A given by

$$p = \begin{pmatrix} 1 & 2 & 3 & 4 & 5 & 6 & 7 & 8 \\ 3 & 6 & 4 & 1 & 7 & 2 & 8 & 5 \end{pmatrix}.$$

Then $p = (1, 3, 4) \circ (2, 6) \circ (5, 7, 8)$. This is because the right-hand side

$$\begin{aligned} &= \begin{pmatrix} 1 & 2 & 3 & 4 & 5 & 6 & 7 & 8 \\ 3 & 2 & 4 & 1 & 5 & 6 & 7 & 8 \end{pmatrix} \circ \begin{pmatrix} 1 & 2 & 3 & 4 & 5 & 6 & 7 & 8 \\ 1 & 6 & 3 & 4 & 5 & 2 & 7 & 8 \end{pmatrix} \\ &\quad \circ \begin{pmatrix} 1 & 2 & 3 & 4 & 5 & 6 & 7 & 8 \\ 1 & 2 & 3 & 4 & 7 & 6 & 8 & 5 \end{pmatrix} \\ &= \begin{pmatrix} 1 & 2 & 3 & 4 & 5 & 6 & 7 & 8 \\ 3 & 2 & 4 & 1 & 5 & 6 & 7 & 8 \end{pmatrix} \circ \begin{pmatrix} 1 & 2 & 3 & 4 & 5 & 6 & 7 & 8 \\ 1 & 6 & 3 & 4 & 7 & 2 & 8 & 5 \end{pmatrix} \\ &= \begin{pmatrix} 1 & 2 & 3 & 4 & 5 & 6 & 7 & 8 \\ 3 & 6 & 4 & 1 & 7 & 2 & 8 & 5 \end{pmatrix} \\ &= p. \end{aligned}$$

Note

1. Consider a permutation p on $A = \{a_1, a_2, \ldots, a_n\}$. To express p as a product of disjoint cycles , we use the following procedure.

   Pick the first element $a_1 \in A$ and write it as the first element of the cycle. Consider $p(a_1)$ and write it as the second element of the cycle. Next consider $p(p(a_1))$ (i.e., the image of the second element under p) and write it as the third element of the cycle. Repeat this until we encounter an element whose image is a_1. Now one cycle is completed. If this cycle contains all the elements of A, then the given permutation itself is a cycle. Otherwise, choose an element of A that has not appeared in the first cycle and write it as the first element of a second cycle. Take the image of this element under p and write it as the second element of the second cycle. Repeat this until we encounter an element whose image under p is the first element of the second cycle. Now the second cycle is completed. Repeat this until all the elements of A are exhausted. Then the permutation p will be the product of all these cycles. The fact that p is one–one and the way these cycles have been generated shows that these cycles are all disjoint.

2. $(a_{i1}, a_{i2}, a_{i3}, \ldots, a_{ik}) = (a_{i2}, a_{i3}, a_{i4}, \ldots, a_{ik}, a_{i1}) = (a_{i3}, a_{i4}, \ldots, a_{ik}, a_{i1}, a_{i2})$, etc.

   In other words, a cycle remains the same if we start with any arbitrary element in the cycle and write the remaining elements of the cycle in the same order in which they appear in the original cycle (after encountering the last element of the original cycle, we write the first element of the original cycle). We can also permute the various cycles in the representation of a permutation as the product of disjoint cycles. In fact, these are the only representations of a permutation as a product of disjoint cycles. A permutation cannot be represented as a product of disjoint cycles in any other manner.

Result Any permutation p can be expressed as a product of disjoint cycles. We can permute the cycles in this representation and obtain other representations of p. In each cycle we can start with any arbitrary element and write the other elements in the same order in which they appear in the original cycle. All of them will give representations of p as a product of disjoint cycles. p cannot be expressed as a product of disjoint cycles in any other manner.

Example Consider the set A and the permutation p given in the above example. The following are some representations of p as a product of disjoint cycles.

$$
\begin{aligned}
p &= (1, 3, 4) \circ (2, 6) \circ (5, 7, 8)\\
&= (1, 3, 4) \circ (5, 7, 8) \circ (2, 6)\\
&= (7, 8, 5) \circ (6, 2) \circ (4, 1, 3)\\
&= (2, 6) \circ (8, 5, 7) \circ (3, 4, 1), \text{ etc.}
\end{aligned}
$$

Any cycle can be represented as a product of transpositions but this representation is not unique. Consider the cycle $(a_{i1}, a_{i2}, \ldots\ldots, a_{ik})$. One method of representing this cycle as a product of transpositions is

$$
\begin{aligned}
(a_{i1}, a_{i2}, \ldots\ldots a_{ik}) &= (a_{i1}, a_{ik}) \circ (a_{i1}, a_{ik-1}) \circ \ldots\ldots \circ (a_{i1}, a_{i2})\\
&= \begin{pmatrix} a_1, & a_2 & a_{i1} & a_{i2} & a_{ik} & a_{j-1} & a_j \\ a_1, & a_2 & a_{ik} & a_{i2} & a_{i1} & a_{j-1} & a_j \end{pmatrix}\\
&\circ \begin{pmatrix} a_1 & a_2 & a_{i1} & a_{i2} & a_{ik-1} & a_{j-1} & a_j \\ a_1 & a_2 & a_{ik-1} & a_{i2} & a_{i1} & a_{j-1} & a_j \end{pmatrix}\\
&\circ \begin{pmatrix} a_1 & a_2 & a_{i1} & a_{i2} & a_{ik-1} & a_{ik} & a_{j-1} & a_j \\ a_1 & a_2 & a_{i2} & a_{i1} & a_{ik-1} & a_{ik} & a_{j-1} & a_j \end{pmatrix}.
\end{aligned}
$$

Example Let $A = \{a_1, a_2, a_3, a_4, a_5\}$. Then,

$$(a_2, a_3, a_4, a_5) = (a_2, a_5) \circ (a_2, a_4) \circ (a_2, a_3).$$

This can be seen from the fact that

$$
\begin{aligned}
\text{R.H.S} &= \begin{pmatrix} a_1 & a_2 & a_3 & a_4 & a_5 \\ a_1 & a_5 & a_3 & a_4 & a_2 \end{pmatrix} \circ \begin{pmatrix} a_1 & a_2 & a_3 & a_4 & a_5 \\ a_1 & a_4 & a_3 & a_2 & a_5 \end{pmatrix}\\
&\circ \begin{pmatrix} a_1 & a_2 & a_3 & a_4 & a_5 \\ a_1 & a_3 & a_2 & a_4 & a_5 \end{pmatrix}\\
&= \begin{pmatrix} a_1 & a_2 & a_3 & a_4 & a_5 \\ a_1 & a_5 & a_3 & a_4 & a_2 \end{pmatrix} \circ \begin{pmatrix} a_1 & a_2 & a_3 & a_4 & a_5 \\ a_1 & a_3 & a_4 & a_2 & a_5 \end{pmatrix}\\
&= \begin{pmatrix} a_1 & a_2 & a_3 & a_4 & a_5 \\ a_1 & a_3 & a_4 & a_5 & a_2 \end{pmatrix}, \quad \text{which is L.H.S.}
\end{aligned}
$$

The above permutation p can be represented as a product of transpositions as shown below.

$$
\begin{aligned}
p &= (1, 3, 4) \circ (2, 6) \circ (5, 7, 8)\\
&= (1, 4) \circ (1, 3) \circ (2, 6) \circ (5, 8) \circ (5, 7).
\end{aligned}
$$

Result Any permutation p can be represented as a product of transpositions but this representation is not unique.

Proof This follows from the fact that any permutation p can be represented as a product of disjoint cycles and any cycle can be represented as a product of transpositions.

Note

1. Though the representation of a permutation p as a product of transpositions is not unique, it can be seen that if in one representation of p as a product of transpositions there are an even number of transpositions, then p cannot be represented as a product of an odd number of transpositions. In other words, in any other representation of p as a product of transpositions, there will be an even number of transpositions. Such a permutation is called an even permutation. Similarly, if a permutation q can be represented as a product of an odd number of transpositions in one representation, then in any other representation of q as a product of transpositions there can be only an odd number of transpositions. Such a permutation is called an odd permutation.
2. If A is a finite set with n elements, then the number of permutations on A is $n!$. This follows from the fact that for the first element of A, its image can be any of the n elements. Hence there are n ways of defining the image of the first element. Since a permutation is a one–one function, the image of the second element can be any one of the remaining $n - 1$ elements. Thus there are $n - 1$ ways of defining the image of the second element. Finally there is only one way of defining the image of the last element. Hence the number of permutations is $n \times (n - 1) \times (n - 2) \times \cdots 3 \times 2 \times 1$, which is nothing but $n!$

   The following theorem says that these $n!$ permutations are equally divided among even and odd permutations.

Theorem Let A be a set consisting of n elements ($n \geq 2$). Then there are $\frac{n!}{2}$ even permutations and $\frac{n!}{2}$ odd permutations.

Proof Let A_n denote the set of all even permutations and B_n denote the set of all odd permutations. Since A contains at least two elements, there is at least one transposition. Fix a transposition q_0. Define a function f from A_n to B_n as $f(p) = q_0 \circ p (p \in A_n)$. Since $p \in A_n$, p is an even permutation, in any representation of p as a product of transpositions there is an even number of transpositions. Hence $q_0 \circ p$ can be represented as a product of an odd number of transpositions proving that $q_0 \circ p \in B_n$. This proves that the function f is well defined. We will prove that f is both one–one and onto.

To prove that f is one–one, assume that $f(p_1) = f(p_2) (p_1, p_2 \in A_n)$. Then $q_0 \circ p_1 = q_0 \circ p_2$ and hence $q_0 \circ (q_0 \circ p_1) = q_0 \circ (q_0 \circ p_2)$. Since $\circ$ is associative, we have $(q_0 \circ q_0) \circ p_1 = (q_0 \circ q_0) \circ p_2$. Since $q_0 \circ q_0 = I_A$, it follows that $p_1 = p_2$ proving that f is one–one.

To prove f is onto, consider $q \in B_n$. Then q is an odd permutation so that $q_0 \circ q$ is an even permutation and hence belongs to A_n. We now have

$$
\begin{aligned}
f(q_0 \circ q) &= q_0 \circ (q_0 \circ q) \\
&= (q_0 \circ q_0) \circ q \qquad \text{(since } \circ \text{ is associative)} \\
&= I_A \circ q = q.
\end{aligned}
$$

This proves that f is onto. We note that A_n and B_n are finite sets and $f\colon A_n \to B_n$ is both one–one and onto. Hence $|A_n| = |B_n|$. Now, $A_n \cup B_n$ is the set of all permutations on A and hence $|A_n \cup B_n| = n!$ Since no permutation can be both even and odd, we have $A_n \cap B_n = \phi$ so that $|A_n \cap B_n| = 0$. Hence $n! = |A_n \cup B_n| = |A_n| + |B_n| = 2|A_n|$ (since $|A_n| = |B_n|$), proving that $|A_n| = \frac{n!}{2} = |B_n|$. This completes the proof of the theorem.

7.5 Characteristic Functions

Let X be a universal set and consider a subset A of X. Define a function $f_A : X \to \{0, 1\}$ as follows.

$$
\begin{aligned}
f_A(x) &= 0 \text{ if } x \notin A. \\
&= 1 \text{ if } x \in A.
\end{aligned}
$$

For any element $x \in X$, there are only two possibilities: either $x \in A$ or $x \notin A$. Hence f_A is a well-defined function. It is called the characteristic function associated with the subset A.

Theorem

(i) $f_{A\cap B} = f_A f_B$

(ii) $f_{A\cup B} = f_A + f_B - f_A f_B$

(iii) $f_{A\oplus B} = f_A + f_B - 2 f_A f_B$

Note The function $f_A f_B$ is defined as $f_A f_B(x) = f_A(x) f_B(x)$ for all $x \in X$, where $f_A(x) f_B(x)$ denotes the product of $f_A(x)$ and $f_B(x)$.

Proof

(i) Consider any element $x \in X$. Then either $x \in A \cap B$ or $x \notin A \cap B$. Consider the case $x \in A \cap B$. Then $f_{A\cap B}(x) = 1$. Since $x \in A \cap B$, $x \in A$

and $x \in B$. Hence $f_A f_B(x) = f_A(x) f_B(x) = 1 \times 1 = 1$, proving that $f_{A \cap B}(x) = f_A f_B(x)$.

Now assume that $x \notin A \cap B$ so that $f_{A \cap B}(x) = 0$. Then either $x \notin A$ or $x \notin B$. This means either $f_A(x) = 0$ or $f_B(x) = 0$ so that $f_A f_B(x) = f_A(x)\ f_B(x) = 0$ again, proving that $f_{A \cap B}(x) = f_A$ $f_B(x)$.

(ii) Consider any element $x \in X$. Then either $x \in A \cup B$ or $x \notin A \cup B$. Consider the case $x \in A \cup B$. Then either $x \in A$ or $x \in B$. We have $f_{A \cup B}(x) = 1$ and

$$\begin{aligned}(f_A + f_B - f_A f_B)(x) &= f_A(x) + f_B(x) - f_A f_B(x)\\ &= f_A(x) + f_B(x) - f_A(x) f_B(x).\end{aligned}$$

If $x \in A$, then the last expression becomes $1 + f_B(x) - f_B(x)$ which is 1. If $x \in B$, the last expression becomes $f_A(x) + 1 - f_A(x)$ which is again 1. This proves that $f_{A \cup B}(x) = (f_A + f_B - f_A f_B)(x)$.

Now consider the case $x \notin A \cup B$. Then $x \notin A$ and $x \notin B$ so that $f_A(x) = 0$ and $f_B(x) = 0$. Now $f_{A \cup B}(x) = 0$ and

$$\begin{aligned}(f_A + f_B - f_A f_B)(x) &= f_A(x) + f_B(x) - f_A f_B(x)\\ &= f_A(x) + f_B(x) - f_A(x) f_B(x) = 0.\end{aligned}$$

(iii) Consider any element $x \in X$. Then, either $x \in A \oplus B$ or $x \notin A \oplus B$. If $x \in A \oplus B$, then $x \in (A - B) \cup (B - A)$ (by definition of $A \oplus B$). Then either $x \in A - B$ or $x \in B - A$. Assume $x \in A - B$. Then $x \in A$ and $x \notin B$. In this case, we have $f_{A \oplus B}(x) = 1$ and $(f_A + f_B - 2 f_A f_B)(x)$ $= f_A(x) + f_B(x) - 2 f_A(x) f_B(x) = 1 + 0 - 0 = 1$. Similarly, if $x \in B - A$, then $(f_A + f_B - 2 f_A f_B)(x) = 1$.

Now assume that $x \notin A \oplus B = (A - B) \cup (B - A)$. Then $x \notin A - B$ and $x \notin B - A$. From $x \notin A - B$ we can conclude $x \notin A$ or $x \in B$. From $x \notin B - A$ we can conclude $x \notin B$ or $x \in A$. Hence the only possibilities are $x \notin A$ and $x \notin B$ or $x \in B$ and $x \in A$. If we consider the possibility $x \notin A$ and $x \notin B$, then

$$\begin{aligned}(f_A + f_B - 2 f_A f_B)(x) &= f_A(x) + f_B(x) - 2 f_A(x) f_B(x)\\ &= 0 + 0 - 0 = 0.\end{aligned}$$

If we consider the possibility $x \in B$ and $x \in A$, then

$$\begin{aligned}(f_A + f_B - 2 f_A f_B)(x) &= f_A(x) + f_B(x) - 2 f_A(x) f_B(x)\\ &= 1 + 1 - 2 \times 1 \times 1 = 0.\end{aligned}$$

Since $x \notin A \oplus B$, we also have $f_{A \oplus B}(x) = 0$. This completes the proof of (iii).

Consider a set S consisting of n elements. Further, assume that the elements of S are arranged in a particular order. Consider any subset A of S. We can associate with A a sequence $(x_1, x_2, \ldots, x_n)$ of 0s and 1s as follows.

If the ith element of S belongs to A, then $x_i = 1$. Otherwise $x_i = 0$. Conversely, consider a sequence $(y_1, y_2, \ldots, y_n)$ of 0s and 1s. We can associate with this sequence a subset B of S as follows.

If $y_i = 1$, then we place ith element of S into B. If $y_i = 0$, then we do not place the ith element of S into B.

Example Consider the set S = {cat, tiger, lion, dog, cow} and the subset A = {tiger, dog, cow}. The sequence associated with A is (0, 1, 0, 1, 1). The subset of S associated with the sequence (1, 0, 0, 1, 0) is {cat, dog}.

Result Let S be a set consisting of n elements and $P(S)$ denote the power set of S. Let T denote the set of all sequences (each of length n) of 0s and 1s. Consider a function $f\colon P(S) \to T$ which associates with each subset A of S a sequence of 0s and 1s as mentioned above. Then the function f is both one–one and onto.

Note Each sequence in T has n components and there are two possible choices for each component (each component can be either 0 or 1). Hence the number of elements in T is 2^n. By the above result, it follows that the number of elements in $P(S)$ is also 2^n.

Exercises

1. Let $A = \{1, 2, 3, 4, 5, 6\}$ and let

$$p_1 = \begin{pmatrix} 1\,2\,3\,4\,5\,6 \\ 3\,4\,1\,2\,6\,5 \end{pmatrix}, \; p_2 = \begin{pmatrix} 1\,2\,3\,4\,5\,6 \\ 6\,3\,2\,5\,4\,1 \end{pmatrix} \text{ and } p_3 = \begin{pmatrix} 1\,2\,3\,4\,5\,6 \\ 2\,3\,1\,5\,4\,6 \end{pmatrix}.$$

   Find p_1^{-1}, $p_3 \circ p_1$, $(p_2 \circ p_1) \circ p_2$, $p_1 \circ (p_3 \circ p_2^{-1})$, $p_3^{-1} \circ p_2^{-1}$, $(p_3 \circ p_2) \circ p_1$, $p_3 \circ (p_2 \circ p_1)^{-1}$.

2. Let $A = \{1, 2, 3, 4, 5, 6, 7, 8\}$. Determine the following.

   (a) $(3, 5, 7, 8) \circ (1, 3, 2)$

   (b) $(2, 6) \circ (3, 5, 7, 8) \circ (2, 5, 3, 4)$

   (c) $(1, 4) \circ (2, 4, 5, 6) \circ (1, 4, 6, 7)$

   (d) $(5, 8) \circ (1, 2, 3, 4) \circ (3, 5, 6, 7)$

3. Let $A = \{1, 2, 3, 4, 5, 6, 7, 8\}$. Write each of the following permutations as a product of disjoint cycles.

(i) $p_1 = \begin{pmatrix} 1 & 2 & 3 & 4 & 5 & 6 & 7 & 8 \\ 4 & 3 & 2 & 5 & 1 & 8 & 7 & 6 \end{pmatrix}$

(ii) $p_2 = \begin{pmatrix} 1 & 2 & 3 & 4 & 5 & 6 & 7 & 8 \\ 6 & 5 & 7 & 8 & 4 & 3 & 2 & 1 \end{pmatrix}$

(iii) $p_3 = \begin{pmatrix} 1 & 2 & 3 & 4 & 5 & 6 & 7 & 8 \\ 2 & 3 & 4 & 1 & 7 & 5 & 8 & 6 \end{pmatrix}$

(iv) $p_4 = \begin{pmatrix} 1 & 2 & 3 & 4 & 5 & 6 & 7 & 8 \\ 2 & 3 & 1 & 4 & 6 & 7 & 8 & 5 \end{pmatrix}$

4. Let $A = \{1, 2, 3, 4, 5, 6, 7, 8\}$. Determine whether the following permutations are even or odd.

(i) $p_1 = \begin{pmatrix} 1 & 2 & 3 & 4 & 5 & 6 & 7 & 8 \\ 4 & 2 & 1 & 6 & 5 & 8 & 7 & 3 \end{pmatrix}$

(ii) $p_2 = \begin{pmatrix} 1 & 2 & 3 & 4 & 5 & 6 & 7 & 8 \\ 7 & 3 & 4 & 2 & 1 & 8 & 6 & 5 \end{pmatrix}$

5. Determine whether the following permutations are even or odd.

(a) $(6, 4, 2, 1, 5)$

(b) $(4, 8) \circ (3, 5, 2, 1) \circ (2, 4, 7, 1)$

8 Lattices and Finite Boolean Algebra

Let $(A, \leq)$ be a poset. An element $a \in A$ is called a *maximal element* if there is no element $b \in A$ such that $a < b$ ($a < b$ means $a \leq b$ and $a \neq b$).

An element $x \in A$ is called a *minimal element* if there is no element $y \in A$ such that $y < x$.

Note In a poset $(A, \leq)$, it is not necessary that every other element should be less than or equal to a maximal element. The only condition is that it cannot be smaller than any other element. Similarly for minimal element.

A poset can have any number of maximal and minimal elements as the following examples show.

Examples

1. Consider the posets whose Hasse diagrams are given below.

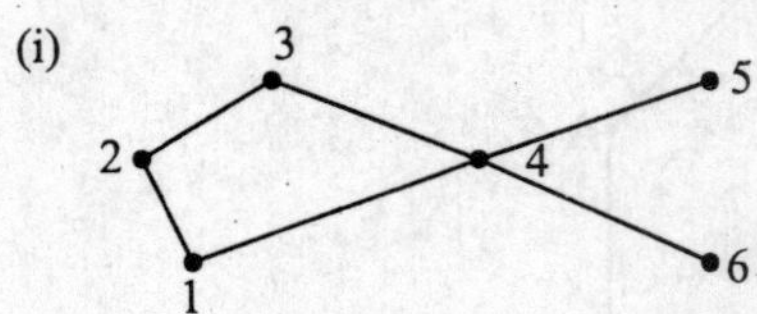

Maximal elements are 3 and 5
Minimal elements are 1 and 6

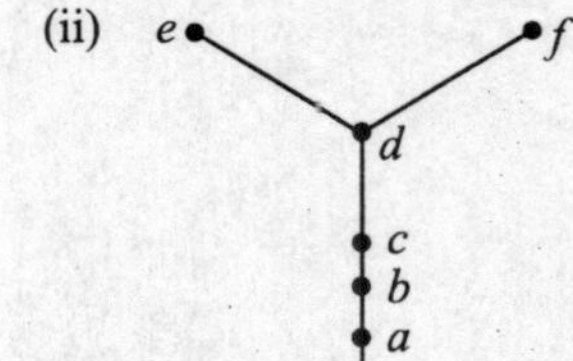

Maximal elements are e and f
Minimal element is a

2. Consider $(\mathbb{R}, \leq)$ where $\mathbb{R}$ denotes the set of all real numbers and $\leq$ denotes usual $\leq$ relation. This poset has no maximal elements and no minimal elements.
3. Let $A = (0, 1] = \{x \in \mathbb{R} | 0 < x \leq 1\}$. Consider the usual $\leq$ relation. 1 is a maximal element of this poset. It has no minimal element.

Theorem Let $(A, \leq)$ be a finite poset. Then A has at least one maximal element and at least one minimal element.

Proof Consider an element $a_1 \in A$. If a_1 is a maximal element, then there is nothing to prove. Otherwise there exists an element $a_2 \in A$ such that $a_1 < a_2$ (by definition of a maximal element). Again, if a_2 is a maximal element, then there is nothing to prove. If a_2 is not a maximal element, then there is an element a_3 such that $a_2 < a_3$. We thus have $a_1 < a_2 < a_3$. Proceeding like this, we obtain $a_1 < a_2 < a_3 < \ldots$. Since A is a finite set, this process has to terminate at some element, say a_k. Then $a_1 < a_2 < a_3 \cdots < a_k$. We cannot find an element b such that $a_k < b$, proving that a_k is a maximal element. Similarly, we can prove that A has at least one minimal element.

Note The above theorem fails for Examples 2 and 3 since $\mathbb{R}$ and A are not finite sets.

Let $(A, \leq)$ be a poset. An element $a \in A$ is called a *greatest element* if $x \leq a$ for all $x \in A$.

An element $b \in A$ is called a *least element* if $b \leq y$ for all $y \in A$.

Examples

1. Consider the poset $(Z^+, \leq)$ where Z^+ denotes the set of all positive integers and $\leq$ denotes the usual less than or equal to relation. This poset has no greatest element. Its least element is zero.
2. For any set S, the poset $(P(S), \subseteq)$ has ϕ as its least element and S as its greatest element.
3. Consider the poset whose Hasse diagram is given below.

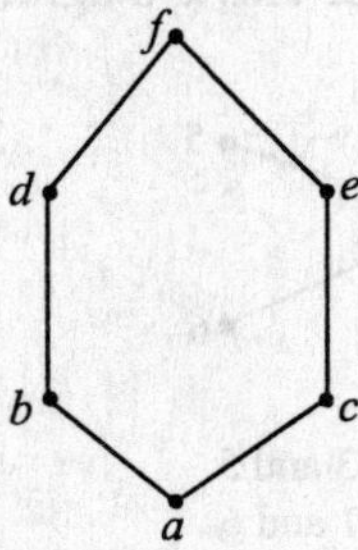

For this poset, the greatest element is f and least element is a.

4. Consider the poset whose Hasse diagram is given below.

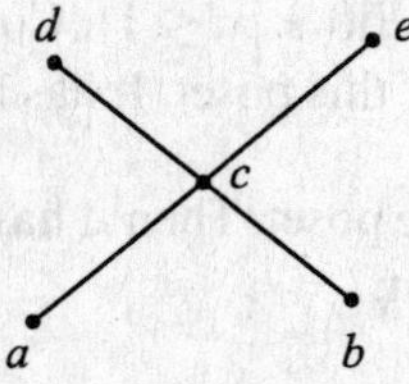

This poset has no greatest element and no least element.

5. Let $A = (0, 1) = \{x \in \mathbb{R} | 0 < x < 1\}$ with the usual less than or equal to relation. This poset has no least element and no greatest element.
6. Let $A = \{2, 4, 6, 8, 12, 18, 24, 36, 72\}$. Define $a \leq b$ if and only if $a|b$ (a divides b). The least element of this poset is 2 whereas the greatest element is 72. Note that 2 divides every element of A and every element of A divides 72.

From the above examples we note that a poset need not have a greatest element. It also need not have a least element. However, we can prove the following.

Result A poset can have at the most one greatest element and at the most one least element .

Proof Suppose a and b are both greatest elements of the poset $(A, \leq)$. Since a is a greatest element, by definition of greatest element it follows that $b \leq a$. Again since b is a greatest element, we have $a \leq b$. Since $\leq$ is antisymmetric, it follows that $a = b$.

Note From the above result we note that if a poset has a greatest element, then it is unique. This greatest element is called the unit element of the poset and is denoted by 1. Similarly, if a poset has a least element, then it is unique. This least element is called zero element and is denoted by 0. 1 and 0 are only symbols. They should not be confused with the numbers 1 and 0.

Let $(A, \leq)$ be a poset and B be a subset of A. An element $a \in A$ is called an *upper bound* of B if $b \leq a$ for all $b \in B$. An element $a \in A$ is called a *least upper bound* (LUB) of B if a is an upper bound of B and $a \leq c$, where c is any upper bound of B. An element $x \in A$ is called a *lower bound* of B if $x \leq y$ for all $y \in B$. An element $x \in A$ is called a *greatest lower bound* (GLB) of B if x is a lower bound of B and $z \leq x$, where z is any lower bound of B.

Note

1. A subset B of a poset $(A, \leq)$ may or may not have a lower bound. B may or may not have an upper bound.
2. An upper bound of B may or may not belong to B. Similarly, a lower bound of B may or may not belong to B.

Examples

1. Consider the poset $(A, \leq)$ where $A = \{a, b, c, d, e, f, g, h\}$ and $\leq$ is given by the following Hasse diagram.

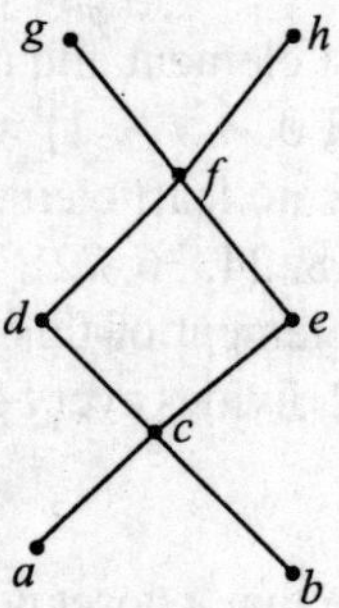

Consider the subset $B = \{c, d, e\}$. Upper bounds of B are f, g and h. LUB is f. Lower bounds of B are a, b and c. GLB is c.

Note that no upper bound of B belongs to B. a and b are lower bounds of B which do not belong to B, but the lower bound c of B belongs to B.

2. Consider the poset given in Example 1 and consider the subset $C = \{b, g, h\}$. C has no upper bounds and hence there is no least upper bound. Lower bound of C is b. Since b is the only lower bound, GLB is also b.
3. Consider the poset $(A, \leq)$ where $A = \{a, b, c, d, e, f\}$ and $\leq$ is given by the following Hasse diagram.

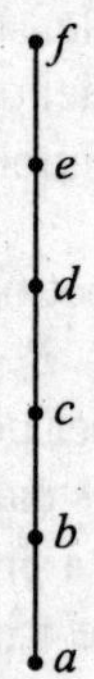

Let $B = \{b, c, d\}$ The upper bounds of B are d, e and f. Lower bounds of B are a and b. The least upper bound of B is d and the greatest lower bound is b.

4. Let $A = R$ and $\leq$ denote the usual less than or equal to relation. Let $B = [1, 2] = \{x \in R | 1 \leq x \leq 2\}$. Lower bounds of $B = \{x \in R | x \leq 1\}$, which is the same as $(-\infty, 1]$. GLB is 1. Upper bounds of $B = \{x \in R | x \geq 2\}$, which is the same as $[2, \infty)$. LUB is 2.
5. Consider the poset $(A, \leq)$ where $A = \{2, 3, 4, 6, 8, 12, 24, 48\}$ and $\leq$ denotes the partial order of divisibility. Consider the subset $B = \{4, 6, 12\}$. Upper bounds of B are 12, 24 and 48. Note that 4 divides 12, 6 divides 12 and 12 divides 12. Hence 12 is an upper bound of B.

Similarly, 24 and 48 are also upper bounds. LUB is 12. 2 is the only lower bound of B. 2 is the only number which divides 4, 6 and 12. Hence GLB is 2.

From the above examples we note that a subset B of a poset $(A, \leq)$ may not have any upper bound or may have one or more than one upper bound. If B has no upper bounds, then obviously B also does not have least upper bound. If B has one or more than one upper bound, then B can have only one least upper bound. Similarly for lower bounds.

Lattice

Let $(A, \leq)$ be a poset. A is said to be a *lattice* if every subset of A consisting of two elements has a least upper bound and a greatest lower bound. In other words, in a lattice any two elements a and b have a least upper bound and a greatest lower bound.

We know the least upper bound of a and b is unique. It is denoted by $a \vee b$ and is called the *join of a and b*. Similarly, the greatest lower bound of a and b which is unique is denoted by $a \wedge b$. It is called the *meet of a and b*.

We have $a \leq a \vee b$ and $b \leq a \vee b$. Whenever $a \leq c$ and $b \leq c$, then $a \vee b \leq c$. Similarly $a \wedge b \leq a$ and $a \wedge b \leq b$. Whenever $d \leq a$ and $d \leq b$, then $d \leq a \wedge b$.

Examples

1. For any set S, $(P(S), \subseteq)$ is a lattice where $A \vee B = A \cup B$ and $A \wedge B = A \cap B (A, B \in P(S))$.
2. For any positive integer n, let D_n denote the set of all positive divisors of n. If $a, b \in D_n$, define $a \leq b$ if and only if $a \mid b$. Then $(D_n, \leq)$ is a lattice where $a \vee b = \text{lcm}(a, b)$ and $a \wedge b = \gcd(a, b)$.
3. Let $A = \{a, b, c, d, e, f, g, h, i\}$ and $\leq$ be given by the Hasse diagram shown below.

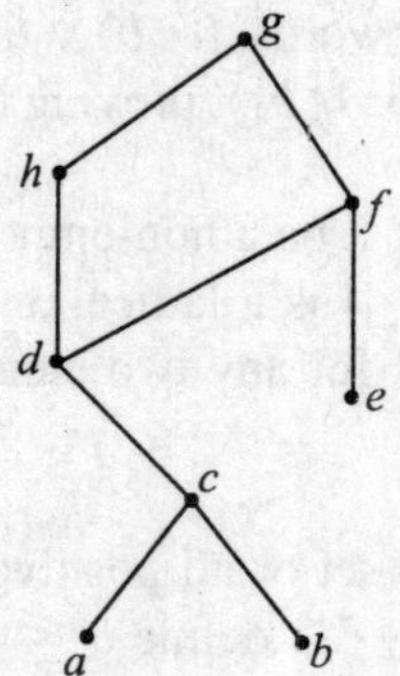

$\{a, b\}$ does not have a lower bound. Hence this poset does not represent a lattice.

Note Whenever we want to prove that $(A, \leq)$ is a lattice, we have to prove that any two elements of A have a least upper bound and a greatest lower bound. Whenever we want to prove that $(A, \leq)$ is not a lattice, we have to find two elements a and b which do not have a least upper bound or a greatest lower bound.

Theorem Let $(A, \leq_a)$ and $(B, \leq_b)$ be two lattices. Define $\leq$ in $A \times B$ as follows. If $(a, b), (a_1, b_1) \in A \times B$, then $(a, b) \leq (a_1, b_1)$ if and only if $a \leq_a a_1$ in A and $b \leq_b b_1$ in B. Then $(A \times B, \leq)$ is a lattice.

Proof We will first prove that $\leq$ is a partial order relation on $A \times B$. For this, we have to prove that $\leq$ is reflexive, antisymmetric and transitive.

If (a, b) is any element of $A \times B (a \in A$ and $b \in B)$, then $a \leq_a a$ and $b \leq_b b$ since $\leq_a$ and $\leq_b$ are reflexive relations. Hence by definition of $\leq$, we have $(a, b) \leq (a, b)$. This proves that $\leq$ is reflexive.

To prove $\leq$ is antisymmetric, assume $(a, b) \leq (a_1, b_1)$ and $(a_1, b_1) \leq (a, b)$, where $a, a_1 \in A$ and $b, b_1 \in B$. By definition of $\leq$, we have $a \leq_a a_1, b \leq_b b_1, a_1 \leq_a a$ and $b_1 \leq_b b$. Since $\leq_a$ is antisymmetric, from $a \leq_a a_1$ and $a_1 \leq_a a$ we conclude that $a = a_1$. Similarly $b = b_1$ so that $(a, b) = (a_1, b_1)$, proving that $\leq$ is antisymmetric.

To prove $\leq$ is transitive, assume that $(a, b) \leq (a_1, b_1)$ and $(a_1, b_1) \leq (a_2, b_2)$. Then $a \leq_a a_1, b \leq_b b_1, a_1 \leq_a a_2$ and $b_1 \leq_b b_2$. Since $\leq_a$ is transitive, $a \leq_a a_1$ and $a_1 \leq_a a_2$ imply $a \leq_a a_2$. Similarly, $b \leq_b b_1$ and $b_1 \leq_b b_2$ imply $b \leq_b b_2$. Hence by definition of $\leq$ we have $(a, b) \leq (a_2, b_2)$, proving that $\leq$ is transitive.

We will now prove that the poset $(A \times B, \leq)$ is a lattice.

Let $(a, b), (a_1, b_1) \in A \times B (a, a_1 \in A$ and $b, b_1 \in B)$. Since $(A, \leq_a)$ is a lattice, a and a_1 have a least upper bound say $a \vee_a a_1$ which belongs to A. Similarly since $(B, \leq_b)$ is a lattice, b and b_1 have a least upper bound, say $b \vee_b b_1$ which belongs to B. Now $(a \vee_a a_1, b \vee_b b_1) \in A \times B$ and it is the least upper bound of (a, b) and (a_1, b_1). In other words, $(a, b) \vee (a_1, b_1) = (a \vee_a a_1, b \vee_b b_1)$ and $(a, b) \wedge (a_1, b_1) = (a \wedge_a a_1, b \wedge_b b_1)$, proving that $(A \times B, \leq)$ is a lattice.

Let $(A, \leq)$ be a lattice and let S be a non-empty subset of A. If $x, y \in S$, then $x, y \in A$ as well. Since $(A, \leq)$ is a lattice, $x \vee y \in A$ and $x \wedge y \in A$. S is said to be a *sublattice* of A if for any two elements $x, y \in S, x \vee y \in S$ and $x \wedge y \in S$.

Example Consider Z^+ (the set of all positive integers) under the relation of divisibility. That is, if $a, b \in Z^+$, define $a \leq b$ if and only if a divides b. If

$m, n \in Z^+$, then $m \vee n = \text{lcm}(m, n)$ and $m \wedge n = \gcd(m, n)$ so that $(Z^+, \leq)$ is a lattice. We also know that for any positive integer n, D_n the set of all positive divisers of n is also a lattice under divisibility. D_n is a sublattice of Z^+.

Definition Let $(A, \leq)$ and $(A_1, \leq_1)$ be two posets and let $f : A \longrightarrow A_1$ be a function which is both one–one and onto. f is said to be an *isomorphism* if whenever $a \leq b$ $(a, b \in A)$, then $f(a) \leq_1 f(b)$. Posets $(A, \leq)$ and $(A_1, \leq_1)$ are said to be *isomorphic* if there exists an isomorphism from A to A_1.

Suppose two posets $(A, \leq)$ and $(A_1, \leq_1)$ are isomorphic. Then there is a function $f : A \to A_1$ which is both one–one and onto. We will prove that $(A, \leq)$ is a lattice if and only if $(A_1, \leq_1)$ is a lattice.

First assume that $(A, \leq)$ is a lattice. We will prove that $(A_1, \leq_1)$ is also a lattice. Let $x, y \in A_1$. We have to prove that $x \vee_1 y$ and $x \wedge_1 y$ both exist, where $x \vee_1 y$ denotes the least upper bound and $x \wedge_1 y$ denotes the greatest lower bound of x and y in A_1. Since f is onto, there exist elements a and b belonging to A such that $f(a) = x$ and $f(b) = y$. Since $(A, \leq)$ is a lattice, $a \vee b \in A$ and $a \wedge b \in A$. We take $x \vee_1 y = f(a \vee b)$ and $x \wedge_1 y = f(a \wedge b)$. Since both $x \vee_1 y$ and $x \wedge_1 y$ exist, it follows that $(A_1, \leq_1)$ is also a lattice. Similarly assuming that $(A_1, \leq_1)$ is a lattice, we can prove that $(A, \leq)$ is also a lattice.

Definition Two lattices $(A, \leq)$ and $(A_1, \leq_1)$ are said to be *isomorphic* if they are isomorphic as posets.

Theorem Let $(A, \leq)$ be a lattice. For any two elements $a, b \in A$, the following are true.

(i) $a \vee b = b$ if and only if $a \leq b$
(ii) $a \wedge b = a$ if and only if $a \leq b$
(iii) $a \wedge b = a$ if and only if $a \vee b = b$

Proof

(i) Suppose $a \vee b = b$. Since $a \leq a \vee b$, it follows that $a \leq b$. Conversely, assume that $a \leq b$. By reflexivity, we have $b \leq b$. Hence b is an upper bound of both a and b. By definition of $a \vee b$ we have $a \vee b \leq b$. Clearly, $b \leq a \vee b$. Since $\leq$ is antisymmetric, it follows that $a \vee b = b$.

(ii) Suppose $a \wedge b = a$. Since $a \wedge b \leq b$, it follows that $a \leq b$. Conversely, assume that $a \leq b$. By reflexivity we have $a \leq a$. Hence a is a lower bound of both a and b. By definition of $a \wedge b$ we have $a \leq a \wedge b$. Clearly, $a \wedge b \leq a$. Since $\leq$ is antisymmetric, it follows that $a \wedge b = a$.

(iii) Suppose $a \wedge b = a$. Then by (ii) we have $a \leq b$ and hence by (i) $a \vee b = b$. Conversely, assume that $a \vee b = b$. Then by (i) $a \leq b$ and hence by (ii) $a \wedge b = a$.

Let L be a linearly ordered set. If $a, b \in L$, we know either $a \leq b$ or $b \leq a$. If $a \leq b$, then by the above result $a \vee b = b$ and $a \wedge b = a$. If $b \leq a$, then $a \vee b = a$ and $a \wedge b = b$. Hence for any two elements $a, b \in L$, both $a \vee b$ and $a \wedge b$ exist, proving that L is a *lattice*. We have thus proved the following.

Result Any linearly ordered set is a lattice.

Properties of a lattice

Let $(A, \leq)$ be a lattice. If $a, b \in A$, then the following properties hold.

1. (a) $a \vee a = a$
   (b) $a \wedge a = a$ (Idempotent property)
2. (a) $a \vee b = b \vee a$
   (b) $a \wedge b = b \wedge a$ (Commutative property)
3. (a) $a \vee (b \vee c) = (a \vee b) \vee c$
   (b) $a \wedge (b \wedge c) = (a \wedge b) \wedge c$ (Associative property)
4. (a) $a \vee (a \wedge b) = a$
   (b) $a \wedge (a \vee b) = a$ (Absorption property)

Proof of 3 (a)

Let $x = a \vee (b \vee c)$ and $y = (a \vee b) \vee c$. We know $a \leq a \vee b$ and $a \vee b \leq (a \vee b) \vee c$. Hence by transitivity we have $a \leq (a \vee b) \vee c$. Again, $b \leq a \vee b$ implies $b \leq (a \vee b) \vee c$ by transitivity. Also $c \leq (a \vee b) \vee c$. Since $b \vee c$ is the least upper bound of b and c, we obtain $b \vee c \leq (a \vee b) \vee c$. This together with $a \leq (a \vee b) \vee c$ gives $a \vee (b \vee c) \leq (a \vee b) \vee c$ since $a \vee (b \vee c)$ is the least upper bound of a and $b \vee c$. We have thus proved that $x \leq y$. Similarly, we can prove that $y \leq x$. Since $\leq$ is antisymmetric, we obtain $x = y$.

Proof of 4 (a)

We have $a \leq a \vee (a \wedge b)$. Also, $a \wedge b \leq a$. By reflexivity, $a \leq a$. Hence a is an upper bound of a and $a \wedge b$. Since $a \vee (a \wedge b)$ is the least upper bound of a and $a \wedge b$, it follows that $a \vee (a \wedge b) \leq a$. This, together with $a \leq a \vee (a \wedge b)$ proves that $a = a \vee (a \wedge b)$.

By property 3 (a), we note that $a \vee (b \vee c) = (a \vee b) \vee c$ can be simply written as $a \vee b \vee c$, omitting the parentheses. Similarly, $a \wedge b \wedge c$ denotes both $(a \wedge b) \wedge c$ and $a \wedge (b \wedge c)$.

Theorem Let S_1 and S_2 be two finite sets containing the same number of elements. Then the lattices $(P(S_1), \subseteq)$ and $(P(S_2, \subseteq)$ are isomorphic and hence their Hasse diagrams are identical.

Proof Let $S_1 = \{x_1, x_2, \ldots, x_n\}$ and $S_2 = \{y_1, y_2, \ldots, y_n\}$. Associate x_1 with y_1, x_2 with $y_2, \ldots, x_n$ with y_n. Define a function $f : P(S_1) \to P(S_2)$ as follows.

If $A = \{x_{i1}, x_{i2}, \ldots, x_{ik}\} \in P(S_1)$, define $f(A) = \{y_{i1}, y_{i2}, \ldots, y_{ik}\}$ ($y_{i1}, y_{i2}, \ldots, y_{ik}$ are elements of S_2 associated with $x_{i1}, x_{i2}, \ldots, x_{ik}$ respectively). Then $f(A)$ is a subset of S_2 and hence $f(A) \in P(S_2)$. We can see that this function is both one–one and onto. Also, if $A, B \in P(S_1)$, then $A \subseteq B$ if and only if $f(A) \subseteq f(B)$. This proves that f is an isomorphism between the posets $(P(S_1), \subseteq)$ and $(P(S_2), \subseteq)$. Hence $(P(S_1), \subseteq)$ and $(P(S_2), \subseteq)$ are isomorphic lattices.

Note From the above theorem it follows that if S is a finite set, then the Hasse diagram of the lattice $(P(S), \subseteq)$ depends only on the number of elements in S and not on the elements of S.

Example Let $S_1 = \{1, 2, 3\}$ and $S_2 = \{x, y, z\}$. The Hasse diagrams of $(P(S_1), \subseteq)$ and $(P(S_2), \subseteq)$ are as follows.

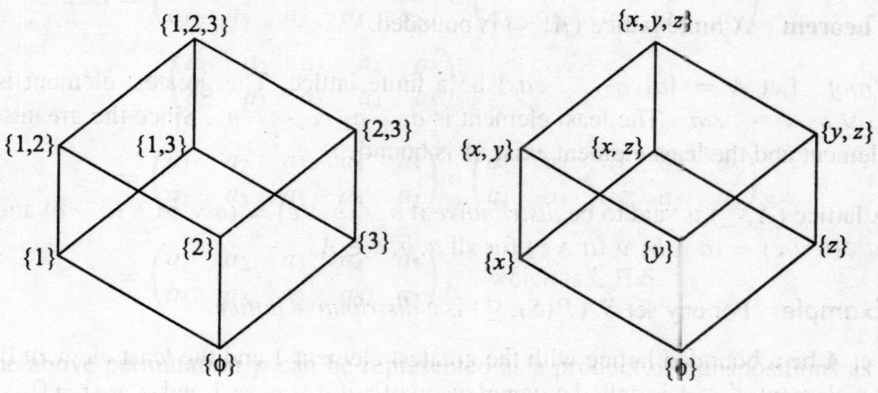

From the above Hasse diagrams we note that the Hasse diagrams are identical but for the labelling of vertices. We also note that the labelling of vertices is not all that important, which means that we can label the vertices with any names we like.

Consider a finite set S consisting of n elements which are arranged in a particular order. We will agree to denote any subset of S by a sequence $x_1, x_2, \ldots, x_n$, where $x_i = 1$ if the ith element of S belongs to the subset and $x_i = 0$ otherwise. In this notation the above Hasse diagrams can be represented by a single Hasse diagram as given below.

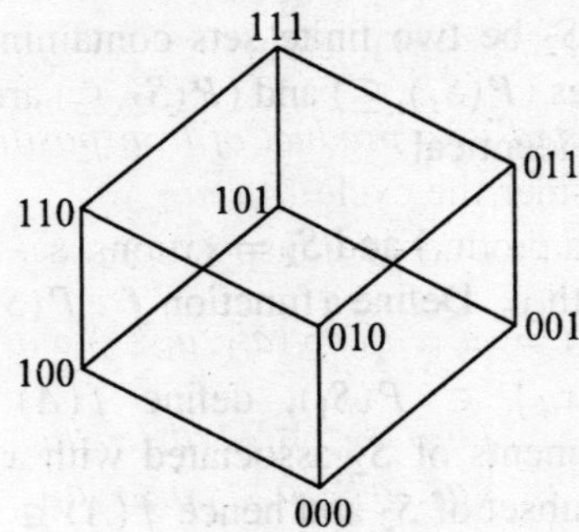

It is not necessary that a lattice should have a greatest element and a least element. However, a lattice A which has a greatest element and a least element is said to be a *bounded lattice*. A bounded lattice can have only one greatest element and only one least element. The greatest element of a bounded lattice is denoted by 1 and the least element by 0.

Examples

1. $(Z, \leq)$ is not a bounded lattice.
2. For any set S, $(P(S), \subseteq)$ is a bounded lattice. The least element is ϕ and the greatest element is S.

Theorem A finite lattice $(A, \leq)$ is bounded.

Proof Let $A = \{a_1, a_2, \ldots, a_n\}$ be a finite lattice. The greatest element is $a_1 \vee a_2 \vee \cdots \vee a_n$. The least element is $a_1 \wedge a_2 \wedge \cdots \wedge a_n$. Since the greatest element and the least element exist, A is bounded.

A lattice $(A, \leq)$ is said to be *distributive* if $a \vee (b \wedge c) = (a \vee b) \wedge (a \vee c)$ and $a \wedge (b \vee c) = (a \wedge b) \vee (a \wedge c)$ for all $a, b, c \in A$.

Example For any set S, $(P(S), \subseteq)$ is a *distributive lattice*.

Let A be a bounded lattice with the greatest element 1 and the least element 0. An element $a' \in A$ is called a *complement* of a if $a \vee a' = 1$ and $a \wedge a' = 0$.

A bounded lattice A is said to be a *complemented lattice* if every element of A has a complement.

Example For any set S, $(P(S), \subseteq)$ is a complemented lattice. If $A \in P(S)$, then A' is nothing but the complement of A (all those elements of S which do not belong to A).

Theorem In a bounded distributive lattice if an element has a complement, it is *unique*.

Proof Let $(A, \leq)$ be a bounded, distributive lattice. Suppose an element $a \in A$ has two complements, say b and c. Then $a \vee b = 1, a \vee c = 1, a \wedge b = 0$ and $a \wedge c = 0$. Now $b = b \vee 0 = b \vee (a \wedge c) = (b \vee a) \wedge (b \vee c) = 1 \wedge (b \vee c) = b \vee c$. Again, $c = c \vee 0 = c \vee (a \wedge b) = (c \vee a) \wedge (c \vee b) = 1 \wedge (c \vee b) = c \vee b = b \vee c = b$.

Let $(A, \leq)$ be a finite poset. $(A, \leq)$ is said to be a *boolean algebra* if $(A, \leq)$ is isomorphic to $(P(S), \subseteq)$ for some finite set S. Since $(P(S), \subseteq)$ is a complemented, distributive lattice, it follows that a finite boolean algebra is also a complemented, distributive lattice.

Example of a boolean algebra

We know $(D_6, \leq)$ is a lattice where $D_6 = \{1, 2, 3, 6\}$ and $\leq$ denotes the relation of divisibility. Its Hasse diagram is as follows.

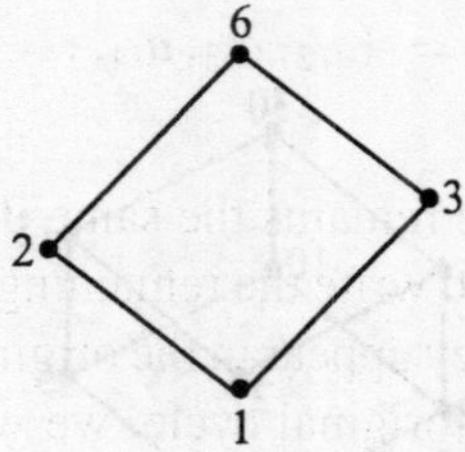

Let S be a set consisting of two elements. Then the Hasse diagram of $(P(S), \subseteq)$ is as follows.

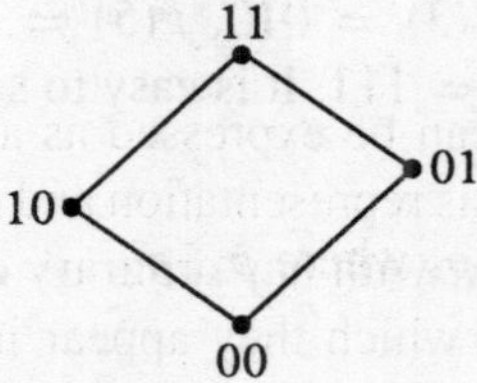

Since the Hasse diagrams are identical, it follows that $(D_6, \leq)$ is isomorphic to $(P(S), \subseteq)$, where S is a set consisting of two elements. Hence $(D_6, \leq)$ is a boolean algebra.

Consider a finite boolean algebra A. Then there is a set S consisting of n elements (for some positive integer n) such that A is isomorphic to $(P(S), \subseteq)$. Since $P(S)$ contains 2^n elements, A also contains 2^n elements. Thus the number of elements in a finite boolean algebra is 2^m for some positive integer m. From this it follows that if in a finite lattice $(A, \leq)$ the number of elements is not equal to 2^k for any positive integer k, then $(A, \leq)$ cannot be a boolean algebra. If the number of elements in A is equal to 2^m for a positive integer m, then A may or may not be a boolean algebra.

Now consider $D_{20} = \{1, 2, 4, 5, 10, 20\}$. Its Hasse diagram is as follows.

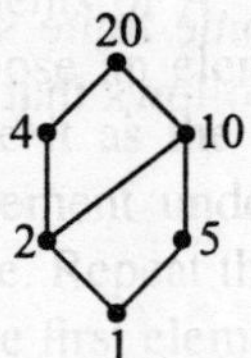

Since the number of elements in D_{20} is 6, which is not equal to 2^k for any positive integer k, it follows that $(D_{20}, \leq)$ is not a boolean algebra.

Now consider $D_{30} = \{1, 2, 3, 5, 6, 10, 15, 30\}$ whose Hasse diagram is given below.

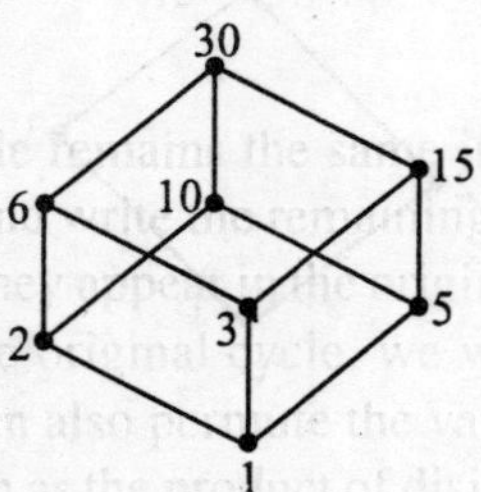

$(D_{30}, \leq)$ is a boolean algebra. D_{30} contains $8 = 2^3$ elements. Taking any set S consisting of 3 elements, we can define a function $f : D_{30} \to P(S)$ as follows. $f(1) = 000$, $f(2) = 100$, $f(3) = 010$, $f(5) = 001$, $f(6) = 110$, $f(10) = 101$, $f(15) = 011$ and $f(30) = 111$. It is easy to see that f is an isomorphism.

Theorem Let $n = p_1 p_2 \ldots p_k$ where $p_1, p_2, \ldots, p_k$ are distinct primes. Then $(D_n, \leq)$ is a boolean algebra.

Proof To prove that D_n is a boolean algebra, we have to identify a finite set S such that $(D_n, \leq)$ is isomorphic to $(P(S), \subseteq)$. Let $S = \{p_1, p_2, \ldots, p_k\}$. We will prove that D_n is isomorphic to $(P(S), \subseteq)$. For this we have to define a function $f : P(S) \to D_n$ which is an isomorphism.

Let $T \in P(S)$. Then T is a subset of S so that T contains some of the prime numbers $p_1, p_2, \ldots, p_k$. Let a_T denote the product of all prime numbers which are in T. Clearly, a_T divides n so that $a_T \in D_n$. Define a function $f : P(S) \to D_n$ as $f(T) = a_T$. We will prove that f is one–one and onto.

If $f(T_1) = f(T_2)$ (T_1 and T_2 are subsets of S), then $a_{T1} = a_{T2}$. This means the product of all prime numbers in T_1 is equal to the product of all prime numbers in T_2, from which it follows that T_1 and T_2 contain the same set of prime numbers. Hence $T_1 = T_2$, proving that f is one–one.

To prove f is onto, consider $d \in D_n$. Then d is a divisor of $n = p_1 p_2 \ldots p_k$ and hence d is a product of some of the prime numbers $p_1, p_2, \ldots, p_k$. If $d = p_{i1} p_{i2} \ldots p_{ij} (p_{i1}, p_{i2}, \ldots, p_{ij} \in \{p_1, p_2, \ldots, p_k\})$, then $d = f(T)$, where $T = \{p_{i1}, p_{i2}, \ldots, p_{ij}\} \subseteq \{p_1, p_2, \ldots, p_k\} = S$ so that $T \in P(S)$.

If $T_1, T_2 \in P(S)$ and $T_1 \subseteq T_2$, then, clearly, the product of all prime numbers in T_1, namely aT_1, divides aT_2 which is the product of all prime numbers in T_2 (since every prime number in T_1 is also in T_2). This proves that $a_{T1} \leq a_{T2}$ and f is an isomorphism of the posets $(P(S), \subseteq)$ and $(D_n, \leq)$.

Result For any two positive integers a and b, $\gcd(a, b) \times \mathrm{lcm}(a, b) = ab$, the product of a and b.

Proof We know that any positive integer can be expressed as a product of primes. Let $p_1, p_2, \ldots, p_l$ be all the prime numbers which appear either in the representation of a as a product of prime numbers or in the representation of b as a product of prime numbers. Then $a = p_1^{a_1} p_2^{a_2} \ldots p_l^{a_l}$ and $b = p_1^{b_1} p_2^{b_2} \ldots p_l^{b_l}$. Note that some of the $a_i s$ and $b_j s$ may be zero. Now

$$\gcd(a, b) = p_1{}^{\min\{a_1, b_1\}} p_2{}^{\min\{a_2, b_2\}} \ldots p_l{}^{\min\{a_l, b_l\}}$$

$$\mathrm{lcm}(a, b) = p_1{}^{\max\{a_1, b_1\}} p_2{}^{\max\{a_2, b_2\}} \ldots p_l{}^{\max\{a_l, b_l\}}$$

$$\begin{aligned}
\gcd(a, b) \times \mathrm{lcm}(a, b) &= p_1{}^{\min\{a_1, b_1\}} p_1{}^{\max\{a_1, b_1\}} p_2{}^{\min\{a_2, b_2\}} p_2{}^{\max\{a_2, b_2\}} \\
&\quad \ldots p_l{}^{\min\{a_l, b_l\}} p_l{}^{\max\{a_l, b_l\}} \\
&= p_1{}^{a_1+b_1} p_2{}^{a_2+b_2} \ldots p_l{}^{a_l+b_l} \\
&= p_1{}^{a_1} p_1{}^{b_1} p_2{}^{a_2} p_2{}^{b_2} \ldots p_l{}^{a_l} p_l{}^{b_l} \\
&= p_1{}^{a_1} p_2{}^{a_2} \ldots p_l{}^{a_l} p_1{}^{b_1} p_2{}^{b_2} \ldots p_l{}^{b_l} \\
&= a \times b
\end{aligned}$$

Note If $\min\{a_i, b_i\} = a_i$, then $\max\{a_i, b_i\} = b_i$ so that $p_i^{\min\{a_i,b_i\}} \times p_i^{\max\{a_i,b_i\}} = p_i^{a_i} \times p_i^{b_i} = p_i^{a_i+b_i}$.

Theorem If p is a prime number such that p^2/n, then $(D_n, \leq)$ is not a boolean algebra.

Proof We know that D_n is bounded with least element 1 and greatest element n. Let, if possible, D_n be a boolean algebra. Since p^2/n, we have $n = p^2q = p(pq)$ so that p/n. Hence $p \in D_n$. Since we assume D_n is a boolean algebra, p has a complement, say p'. By definition of complement, we have $p \wedge p' = 1$ and $p \vee p' = n$. In D_n, the greatest lower bound of any two elements is their GCD and the least upper bound is their LCM. Hence gcd $(p, p') = p \wedge p' = 1$ and lcm $(p, p') = p \vee p' = n$. By the above result we have gcd $(p, p') \times$ lcm $(p, p') = pp' = 1 \times n = n$, proving that $p' = n/p = pq$. Substituting $p' = pq$ in gcd $(p, p') = 1$, we obtain gcd $(p, pq) = 1$ which is a contradiction since p is a common divisor of p and pq. This contradiction proves that D_n is not a boolean algebra.

Since a boolean algebra is also a complemented distributive lattice, it can be represented as $(A, \vee, \wedge, ', 0, 1)$ where A is a set, $\vee$ and $\wedge$ denote the least upper bound and the greatest lower bound respectively, $'$ represents complement, 0 and 1 represent the least element and the greatest element respectively.

Consider the boolean algebra $(B, \vee, \wedge, ', 0, 1)$. Let S be a subset of B. Suppose $0 \in S$, $1 \in S$, $a' \in S$ for all $a \in S$, $a \vee b \in S$ and $a \wedge b \in S$ for all $a, b \in S$. Then S is called a sub-boolean algebra. In other words, a sub-boolean algebra is simply a subset of a boolean algebra which contains the least element and the greatest element of a boolean algebra and which is closed under complementation, meet and join operations of the boolean algebra.

Example Let $B_n = \{(x_1, x_2, x_3, \ldots, x_n) | x_i = 0 \text{ or } 1\}$. B_n is thus the set of all n tuples of 0 s and 1 s. If $x = (x_1, x_2, x_3, \ldots, x_n)$ and $y = (y_1, y_2, y_3, \ldots, y_n) \in B_n$, define

$$x \wedge y = (x_1 \wedge y_1, x_2 \wedge y_2, x_3 \wedge y_3, \ldots, x_n \wedge y_n),$$

$$x \vee y = (x_1 \vee y_1, x_2 \vee y_2, x_3 \vee y_3, \ldots, x_n \vee y_n),$$

$$x' = (x_1', x_2', x_3', \ldots, x_n'),$$

$$0_n = (0, 0, 0, \ldots, 0),$$

$$1_n = (1, 1, 1, \ldots, 1).$$

Here $x_i \wedge y_i = 1$ if and only if $x_i = y_i = 1$ and $x_i \vee y_i = 0$ if and only if $x_i = y_i = 0$. $x'_k = 1$ if and only if $x_k = 0$.

Consider two boolean algebras $(B_1, \wedge_1, \vee_1, ', 0_1, 1_1)$ and $(B_2\ \wedge_2, \vee_2, ''\ 0_2, 1_2)$. Then $(B_1 \times B_2, \wedge_3, \vee_3, ''', 0_3, 1_3)$ is a boolean algebra where $\wedge_3, \vee_3, '''$, 0_3 and 1_3 are defined as follows.

If $a, a_1 \in B_1$ and $b, b_1 \in B_2$, then,

$$(a, b) \wedge_3 (a_1, b_1) = (a \wedge_1 a_1, b \wedge_2 b_1),$$

$$(a, b) \vee_3 (a_1, b_1) = (a \vee_1 a_1, b \vee_2 b_1),$$

$$(a, b)''' = (a', b''),$$

$$0_3 = (0_1, 0_2),$$

$$1_3 = (1_1, 1_2).$$

A function $f : B_1 \to B_2$ is said to be a boolean homomorphism if the following conditions are satisfied.

$$f(a \wedge_1 b) = f(a) \wedge_2 f(b)$$

$$f(a \vee_1 b) = f(a) \vee_2 f(b)$$

$$f(a') = (f(a))''$$

$$f(0_1) = 0_2$$

$$f(1_1) = 1_2$$

A boolean expression in n variables $x_1, x_2, x_3, \ldots, x_n$ is defined recursively as follows.

(i) 0 and 1 are boolean expressions.
(ii) $x_1, x_2, x_3, \ldots, x_n$ are boolean expressions.
(iii) If α and β are boolean expressions, then $\alpha \vee \beta$, $\alpha \wedge \beta$ and α' are also boolean expressions.
(iv) Nothing else is a boolean expression.

Rule (iv) says that any boolean expression is formed using only rules (i) to (iii).

Solved problems

1. For elements a, b, c in a boolean algebra, prove that $b \wedge (a \vee (a' \wedge (b \vee b'))) = b$.

Proof We have $b \wedge (a \vee (a' \wedge (b \vee b'))) = b \wedge (a \vee (a' \wedge 1))$
$= b \wedge (a \vee a') = b \wedge 1 = b.$

2. If $B = \{0, 1\}$, compute the truth table of the function $f : B_3 \to B$ defined by p in each of the following cases.

(i) $p(x, y, z) = x \wedge (y \vee \sim z)$.

Solution

x	y	z	$\sim z$	$y \vee \sim z$	$x \wedge (y \vee \sim z)$
1	1	1	0	1	1
1	1	0	1	1	1
1	0	1	0	0	0
0	1	1	0	1	0
0	0	1	0	0	0
0	1	0	1	1	0
1	0	0	1	1	1
0	0	0	1	1	0

(ii) $p(x, y, z) = (x \vee y) \wedge (z \vee \sim x)$

Solution

x	y	z	$\sim x$	$x \vee y$	$z \vee \sim x$	$(x \vee y) \wedge (z \vee \sim x)$
1	1	1	0	1	1	1
1	1	0	0	1	0	0
1	0	1	0	1	1	1
0	1	1	1	1	1	1
0	0	1	1	0	1	0
0	1	0	1	1	1	1
1	0	0	0	1	0	0
0	0	0	1	0	0	0

(iii) $p(x, y, z) = (x \wedge \sim y) \vee (y \wedge (\sim x \vee y))$.

Solution

x	y	z	$(x \wedge \sim y)$	$(\sim x \vee y)$	$y \wedge (\sim x \vee y)$	$(x \wedge \sim y) \vee$ $(y \wedge (\sim x \vee y))$
1	1	1	0	1	1	1
1	1	0	0	1	1	1
1	0	1	1	0	0	1
0	1	1	0	1	1	1
0	0	1	0	1	0	0
0	1	0	0	1	1	1
1	0	0	1	0	0	1
0	0	0	0	1	0	0

(iv) $p(x, y, z) = (x \wedge y) \vee (\sim x \wedge (y \wedge \sim z))$

Solution

x	y	z	$\sim x$	$\sim z$	$(x \wedge y)$	$(y \wedge \sim z)$	$(\sim x \wedge (y \wedge \sim z))$	Result
1	1	1	0	0	1	0	0	1
1	1	0	0	1	1	0	0	1
1	0	1	0	0	0	0	0	0
0	1	1	1	0	0	0	0	0
0	0	1	1	0	0	0	0	0
0	1	0	1	1	0	1	1	1
1	0	0	0	1	0	0	0	0
0	0	0	0	0	0	0	0	0

3. In the following problems, prove that the given boolean polynomials are equivalent.

(i) $(x \vee y) \wedge (\sim x \vee y)$ and y.

Solution We have $(x \vee y) \wedge (\sim x \vee y)$

$\equiv (x \wedge \sim x) \vee y$ (by distributivity)

$\equiv 0 \vee y$ $\quad$ ($x \wedge \sim x$ is false or 0).

$\equiv y$

(ii) $x \wedge (y \vee (\sim y \wedge (y \vee \sim y)))$ and x.

Solution We have $x \wedge (y \vee (\sim y \wedge (y \vee \sim y)))$

$\equiv x \wedge (y \vee ((\sim y \wedge y) \vee \sim y))$

(by distributivity and $\sim y \wedge \sim y \equiv \sim y$)

$\equiv x \wedge (y \vee 0 \vee \sim y)$(since $\sim y \wedge y \equiv 0$)

$\equiv x \wedge 1$ (since $y \vee 0 \equiv y$ and $y \vee \sim y \equiv 1$)

$\equiv x.$

(iii) $(\sim z \vee x) \wedge ((x \wedge y) \vee z) \wedge (\sim z \vee y)$ and $x \wedge y$.

Solution We have $(\sim z \vee x) \wedge ((x \wedge y) \vee z) \wedge (\sim z \vee y)$

$\equiv ((\sim z \vee x) \wedge (x \vee z)) \wedge ((y \vee z) \wedge (\sim z \vee y))$

(by distributivity)

$\equiv (x \vee (z \wedge \sim z)) \wedge y \vee ((z \wedge \sim z))$ (by distributivity)

$\equiv (x \vee 0) \wedge (y \vee 0)$(since $z \wedge \sim z \equiv 0$)

$\equiv x \wedge y$ (since $x \vee 0 \equiv x$ and $y \vee 0 \equiv y$).

Exercises

1. A lattice $(A, \leq)$ is said to be modular if for $a, b, c \in A$, whenever $a \leq c$, we have $a \vee (b \wedge c) = (a \vee b) \wedge c$. Prove that a distributive lattice is modular.
2. Show that the lattice whose diagram is given below is modular but not distributive.

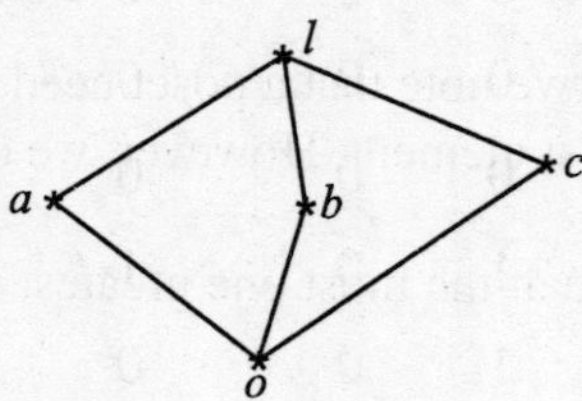

3. Determine whether the following posets are boolean algebras.

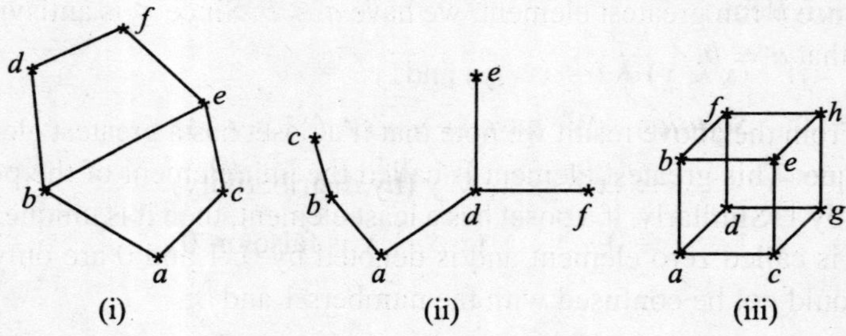

4. For elements a, b, c in a boolean algebra, prove the following.
   (i) $(a \wedge b) \vee (a \wedge b') = a$
   (ii) $(a \wedge b \wedge c) \vee (b \wedge c) = b \wedge c$
   (iii) $((a \vee c) \wedge (b' \vee c))' = (a' \vee b) \wedge c'$
   (iv) $a \vee (b \wedge c) = b \wedge (a \vee c)$ if $a \leq b$.

Karnaugh maps

For two variables x and y, consider the boolean expressions $x \vee y$, $x \wedge y$ and x'. These boolean expressions can be pictorially shown as follows.

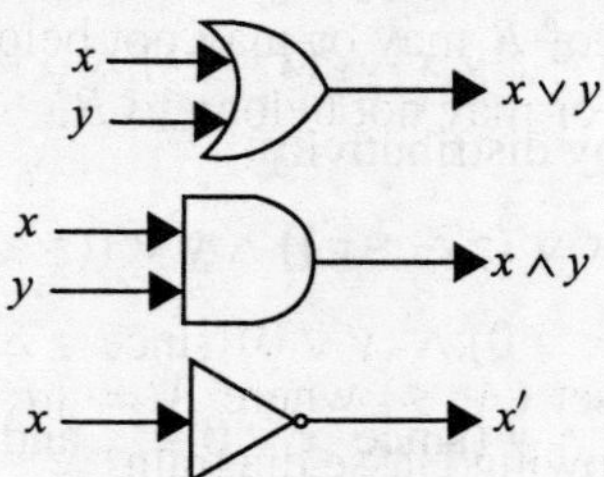

Note that each diagram has lines for the variables on the left and a line on the right which represents the expression as a whole. The diagram for $x \vee y$ is called an OR gate, the diagram for $x \wedge y$ is called an AND gate and the diagram for x' is called an inverter. Since a boolean expression is formed using variables, $\vee$, $\wedge$ and $'$, by repeatedly substituting the above diagrams for $\vee$, $\wedge$ and $'$ it is possible to construct a diagram (called logic diagram) which represents any boolean expression. This is illustrated in the following example.

Example Consider the boolean expression $p(x, y, z) = (x \wedge y) \vee (y \wedge z')$. Its logic diagram is as follows.

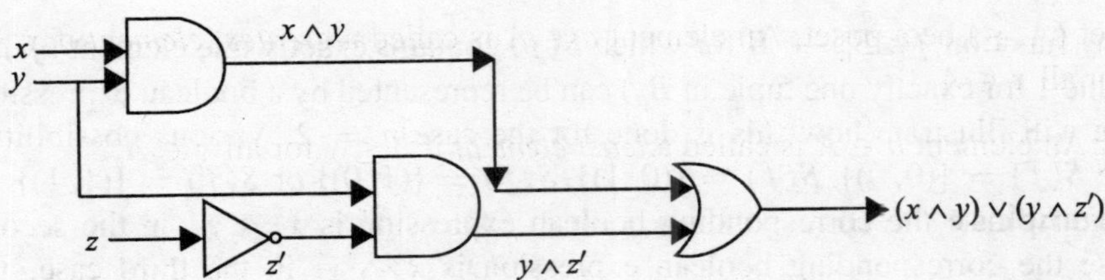

We will now show that any function $f : B_n \to B$ can be produced by a boolean expression. $S(f)$ denotes the set of all elements of B_n which are associated with 1 by the function f. In other words, $S(f) = \{b \in B_n \,|\, f(b) = 1\}$. We will first prove the following.

Theorem Let $f : B_n \to B$, $f_1 : B_n \to B$ and $f_2 : B_n \to B$ be three functions.

(i) If $S(f) = S(f_1) \cup S(f_2)$, then for all $b \in B_n$, $f(b) = f_1(b) \vee f_2(b)$.
(ii) If $S(f) = S(f_1) \cap S(f_2)$, then for all $b \in B_n$, $f(b) = f_1(b) \wedge f_2(b)$.

Proof We will prove only (i). (ii) can be proved in a similar manner. Suppose $b \in S(f)$. Then by definition of $S(f)$, $f(b) = 1$. Since $S(f) = S(f_1) \cup S(f_2)$, either $b \in S(f_1)$, in which case $f_1(b) = 1$ or $b \in S(f_2)$, in which case $f_2(b) = 1$. Hence $f_1(b) \vee f_2(b) = 1$, proving that $f(b) = f_1(b) \vee f_2(b)$.

Now assume that $b \notin S(f)$. Then by definition of $S(f)$, $f(b) = 0$. Since $S(f) = S(f_1) \cup S(f_2)$, $b \notin S(f_1)$ and $b \notin S(f_2)$. Thus $f_1(b) = 0$ and $f_2(b) = 0$. Hence $f_1(b) \vee f_2(b) = 0$, again proving that $f(b) = f_1(b) \vee f_2(b)$.

Example Let $f : B_2 \to B$, $f_1 : B_2 \to B$ and $f_2 : B_2 \to B$ be the functions given below.

x	y	$f_1(x, y)$	x	y	$f_2(x, y)$	x	y	$f(x, y)$
0	0	1	0	0	1	0	0	1
0	1	1	0	1	0	0	1	1
1	0	0	1	0	1	1	0	1
1	1	0	1	1	0	1	1	0

We can easily see that f_1 is represented by the boolean expression x' (note that when x is 0, $f_1(x, y)$ is 1 and when x is 1, $f_1(x, y)$ is 0), f_2 is represented by the boolean expression y' (note that when y is 0, $f_2(x, y)$ is 1 and when y is 1, $f_2(x, y)$ is 0) and $S(f) = S(f_1) \cup S(f_2)$. Hence by the above theorem the boolean expression corresponding to f is $x' \vee y'$.

Any function $f : B_n \to B$ for which $S(f)$ contains exactly one element (f has value 1 for exactly one tuple in B_n) can be represented by a boolean expression. We will illustrate how this is done for the case $n = 2$. Various possibilities are $S(f) = \{(0, 0)\}$, $S(f) = \{(0, 1)\}$, $S(f) = \{(1, 0)\}$ or $S(f) = \{(1, 1)\}$. In the first case the corresponding boolean expression is $x' \wedge y'$; in the second case the corresponding boolean expression is $x' \wedge y$; in the third case, the corresponding boolean expression is $x \wedge y'$; and in the last case the corresponding boolean expression is $x \wedge y$. In general, if $S(f)$ contains the only element $b = (x_1, x_2, \ldots, x_n)$ of B_n, then if $x_i = 1$, we take the ith variable as it is. If $x_j = 0$, we take the negation of jth variable. The conjunction of all of them called the min term gives the corresponding boolean expression. We will denote it by E_b. We will actually prove that this result is true even if $S(f)$ contains more than one element.

Theorem Any function $f : B_n \to B$ can be represented by a boolean expression.

Proof Let $b_1, b_2, \ldots, b_k$ be all the elements of $S(f)$. Note that each of $b_1, b_2, \ldots, b_k$ belongs to B_n. For each i, define a function $f_i : B_n \to B$ as follows.

$$f_i(b_i) = 1$$
$$f_i(b) = 0 \text{ for all } b \neq b_i.$$

Thus f_i has value 1 only for b_i and hence $S(f_i) = \{b_i\}$. Then $S(f) = \{b_1, b_2, \ldots, b_k\} = S(f_1) \cup S(f_2) \cup \cdots \cup S(f_n)$ and hence by the previous theorem $f = f_1 \vee f_2 \vee \cdots \vee f_n$. Since $S(f_i)$ contains only b_i, we know that f_i can be represented by a boolean expression E_{bi}. Hence $f = E_{b_1} \vee E_{b2} \vee \cdots \vee E_{bn}$, proving that f is represented by a boolean expression. We thus see that the boolean expression corresponding to a function $f : B_n \to B$ is obtained by taking "OR" combination of the min terms.

We will now illustrate the above method by means of examples.

Examples Find the boolean expressions for each of the following functions.

(i)

x	y	$f(x, y)$
0	0	1
0	1	0
1	0	0
1	1	1

Answer: $(x' \wedge y') \vee (x \wedge y)$.

(ii)

x	y	z	$f(x, y, z)$
0	0	0	1
0	0	1	1
0	1	0	0
0	1	1	0
1	0	0	1
1	0	1	0
1	1	0	1
1	1	1	0

Answer: $(x' \wedge y' \wedge z') \vee (x' \wedge y' \wedge z) \vee (x \wedge y' \wedge z') \vee (x \wedge y \wedge z')$

$$\equiv [(x' \wedge y' \wedge z') \vee (x' \wedge y' \wedge z)] \vee [(x \wedge y' \wedge z') \vee (x \wedge y \wedge z')]$$

$$\equiv [(x' \wedge y') \wedge (z' \vee z)] \vee [(x \wedge z') \wedge (y' \vee y)]$$

$$\equiv (x' \wedge y') \vee (x \wedge z')$$

We will now see how Karnaugh maps can be used to represent functions $f : B_n \to B$ and to obtain the corresponding boolean expressions. We will assume n has only three possible values 2, 3 or 4.

First assume that n has the value 2. Hence f is a function of two variables, say x and y. If $b \in B_2$, what are the possible values of b ? (0, 0), (0, 1), (1, 0) and (1, 1). We enter them into a 2×2 matrix as shown below.

	y'	y
x'	00	01
x	10	11

In the following figure, we will replace each of these values with the corresponding min term

	y'	y
x'	$x' \wedge y'$	$x' \wedge y$
x	$x \wedge y'$	$x \wedge y$

Note that in the first row, x variable appears everywhere as x'. Hence we have labelled the first row as x'. Similarly, second row and the two columns.

Consider the function $f : B_2 \to B$ whose truth table is given below.

x	y	$f(x, y)$
0	0	1
0	1	1
1	0	0
1	1	0

We will arrange the values of f in the appropriate squares. The resulting 2×2 array of 0s and 1s is called the Karnaugh map of f.

	y'	y
x'	1	1
x	0	0

Now $S(f) = \{(0, 0), (0, 1)\}$ and hence the boolean expression corresponding to f is $(x' \wedge y') \vee (x' \wedge y) = x' \wedge (y' \vee y) = x'$.

The output x' can be obtained just by looking at the Karnaugh map. Note that the row labelled x' is full of 1s. Hence the boolean expression corresponding to f is x'. In general, if one row or one column contains only 1s, then the label of that row or column gives the corresponding boolean expression. If only one cell contains 1 (note that this means $S(f)$ contains only one element), then the corresponding min term gives the boolean expression. What will happen if a row and a column is full of 1s ?

Consider the function $f : B_2 \to B$ whose truth table is given below.

x	y	$f(x, y)$
0	0	1
0	1	1
1	0	1
1	1	0

The Karnaugh map of f is as follows.

	y'	y
x'	1	1
x	1	0

We have decomposed the 1-values into two rectangles. What is the boolean expression for the function having 1s in the horizontal rectangle ? It is x'. The boolean expression for the function having 1s in the vertical rectangle is y'. Hence the boolean expression corresponding to f is $x' \vee y'$.

Suppose we decompose the 1-values into rectangles as shown below.

	y'	y
x'	1	1
x	1	0

Corresponding to this decomposition, we obtain the boolean expression $y' \vee (x' \wedge y)$. Note that the column labelled y' is full of 1s. Hence we have y'. The cell corresponding to row labelled x' and column labelled y contains 1. Hence we have $x' \wedge y$.

We will now consider the case $n = 3$. We thus have a function $f : B_3 \to B$ which is a function of three variables, say x, y and z. We know B_3 has 8 elements. Instead of arranging these 8 elements in a cube, we will use a rectangle of size 2×4 as shown below.

	0 0	0 1	1 1	1 0
0	0 0 0	0 0 1	0 1 1	0 1 0
1	1 0 0	1 0 1	1 1 1	1 1 0

In the following figure, we replace the entries of the cells by the corresponding min terms. Note that the place where 0 appears is replaced by the negation of the corresponding variable and the place where 1 appears is replaced by the corresponding variable itself.

	y'	y'	y	y
x'	$x' \wedge y' \wedge z'$	$x' \wedge y' \wedge z$	$x' \wedge y \wedge z$	$x' \wedge y \wedge z'$
x	$x \wedge y' \wedge z'$	$x \wedge y' \wedge z$	$x \wedge y \wedge z$	$x \wedge y \wedge z'$
	z'	z	z	z'

Consider the shaded regions shown below.

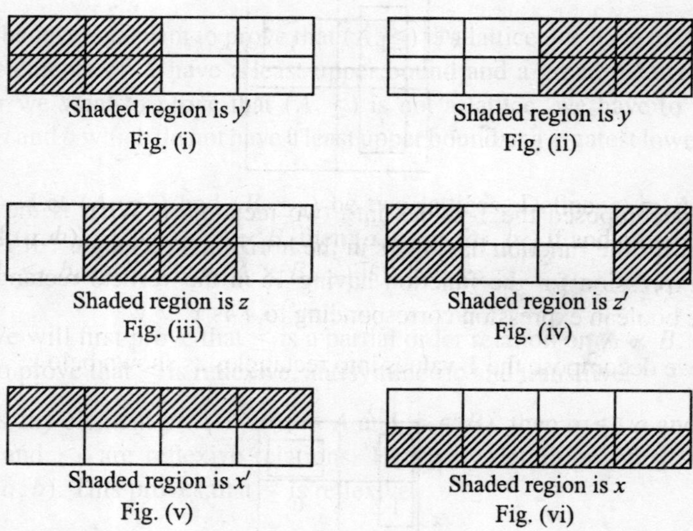

Shaded region is y'
Fig. (i)

Shaded region is y
Fig. (ii)

Shaded region is z
Fig. (iii)

Shaded region is z'
Fig. (iv)

Shaded region is x'
Fig. (v)

Shaded region is x
Fig. (vi)

Suppose 1 appears in the shaded regions of figure (iv). We know the function f can be computed by ORing all the min terms which appear in these shaded regions. In other words, the corresponding boolean expression becomes

$$(x' \wedge y' \wedge z') \vee (x' \wedge y \wedge z') \vee (x \wedge y' \wedge z') \vee (x \wedge y \wedge z')$$

$$\equiv [(x' \wedge y' \wedge z') \vee (x' \wedge y \wedge z')] \vee [(x \wedge y' \wedge z') \vee (x \wedge y \wedge z')]$$

$$\equiv [(x' \wedge z') \wedge (y' \vee y)] \vee [(x \wedge z') \wedge (y' \vee y)]$$

$$\equiv (x' \wedge z') \vee (x \wedge z')$$

$$\equiv (x' \vee x) \wedge z'$$

$$\equiv z'.$$

Similarly, we can see that if 1 appears in the shaded regions of figure (i), then the corresponding boolean expression is y'; if 1 appears in the shaded regions of figure (ii), then the corresponding boolean expression is y; if 1 appears in the shaded regions of figure (iii), then the corresponding boolean expression is z; if 1 appears in the shaded regions of figure (v), then the corresponding boolean expression is x'; and if 1 appears in the shaded regions of figure (vi), then the corresponding boolean expression is x. It is sufficient to consider the boolean expression corresponding to these six basic regions.

Suppose 1 occurs in the shaded region given below.

0	0	1	0
0	0	1	0

We note that this shaded region is obtained by taking the intersection of the shaded regions of the figures given in (ii) and (iii). Hence the corresponding boolean expression is $y \wedge z$ (note that the boolean expression corresponding to (ii) is y whereas the boolean expression corresponding to (iii) is z). Similarly, we can obtain the boolean expression if 1 occurs in any of the other three columns.

Suppose 1 occurs in the shaded region given below.

0	0	0	0
0	0	1	1

This shaded region is obtained by taking intersection of the shaded regions of the figures given in figures (ii) and (vi). Hence the corresponding boolean expression is $x \wedge y$ (note that the boolean expression corresponding to (ii) is y whereas the one corresponding to (vi) is x). Similarly, we can obtain the boolean expression if 1 occurs in any two horizontally adjacent squares.

Suppose 1 occurs in the shaded region given below.

0	0	0	0
1	0	0	1

This shaded region is obtained by taking intersection of the shaded regions of the figures given in figures (iv) and (vi). Hence the corresponding boolean expression is $x \wedge z'$ (note that the boolean expression corresponding to (iv) is z' whereas the boolean expression corresponding to (vi) is x).

Suppose we intersect three of the basic regions and the intersection is not empty. Then the intersection must be a single square. Hence the resulting boolean expression is nothing but the corresponding min term. For example, suppose 1 occurs in the shaded region given below.

0	0	0	1
0	0	0	0

The corresponding boolean expression is $x' \wedge y \wedge z'$.

Consider the function f whose truth table is given below.

x	y	z	$f(x, y, z)$
0	0	0	1
0	0	1	1
0	1	0	0
0	1	1	1
1	0	0	1
1	0	1	0
1	1	0	0
1	1	1	1

The Karnaugh map of f is as follows.

1	1	1	0
1	0	1	0

One decomposition of 1-values of f is also shown in the above figure. A boolean expression for f is $(y' \wedge z') \vee (x' \wedge y') \vee (y \wedge z)$.

We will now consider a function $f : B_4 \to B$ which is a function of four variables x, y, z and w. We will arrange the sixteen elements of B_4 as shown below.

	00	01	11	10
00	0000	0001	0011	0010
01	0100	0101	0111	0110
11	1100	1101	1111	1110
10	1000	1001	1011	1010

We assume that the first and last columns are adjacent as also are the first and last rows. As in the case of a function of 3 variables, the various basic regions are given below.

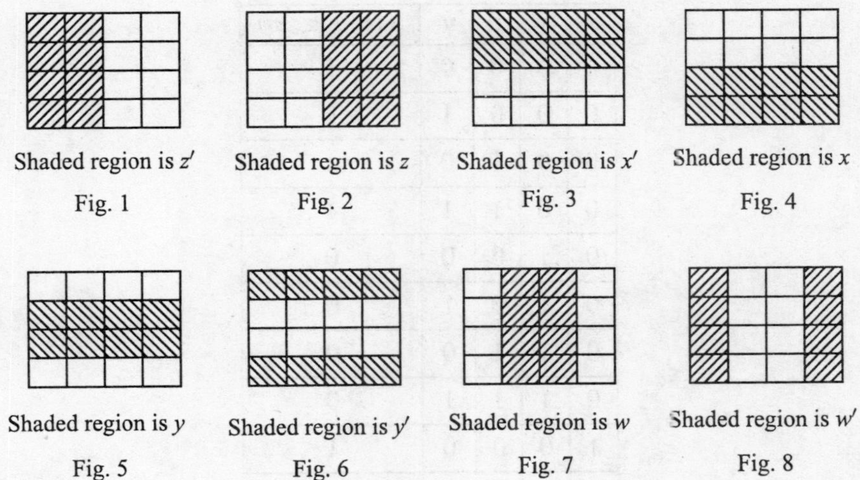

Shaded region is z' — Fig. 1; Shaded region is z — Fig. 2; Shaded region is x' — Fig. 3; Shaded region is x — Fig. 4

Shaded region is y — Fig. 5; Shaded region is y' — Fig. 6; Shaded region is w — Fig. 7; Shaded region is w' — Fig. 8

Consider Fig. 6. Suppose 1 occurs only in the shaded regions. What is the corresponding boolean expression ? It is

$$[(x' \wedge y' \wedge z' \wedge w') \vee (x' \wedge y' \wedge z' \wedge w)] \vee [(x' \wedge y' \wedge z \wedge w)$$
$$\vee (x' \wedge y' \wedge z \wedge w')] \vee$$
$$[(x \wedge y' \wedge z' \wedge w') \vee (x \wedge y' \wedge z' \wedge w)] \vee [(x \wedge y' \wedge z \wedge w)$$
$$\vee (x \wedge y' \wedge z \wedge w')]$$
$$\equiv [(x' \wedge y' \wedge z') \wedge (w' \vee w)] \vee [(x' \wedge y' \wedge z) \wedge (w \vee w')]$$
$$\vee (x \wedge y' \wedge z') \wedge (w \vee w')] \vee [(x \wedge y' \wedge z) \wedge (w \vee w')]$$
$$\equiv (x' \wedge y' \wedge z') \vee (x' \wedge y' \wedge z) \vee (x \wedge y' \wedge z') \vee (x \wedge y' \wedge z)$$
$$\equiv [(x' \wedge y') \wedge (z' \vee z)] \vee [(x \wedge y') \wedge (z' \vee z)]$$
$$\equiv (x' \wedge y') \vee (x \wedge y')$$
$$\equiv (x' \vee x) \wedge y'$$
$$\equiv y'.$$

Similarly, we can see that the boolean expressions corresponding to Fig. 1 is z'; Fig. 2 is z; Fig. 3 is x'; Fig. 4 is x; Fig. 5 is y; Fig. 7 is w; and finally the boolean expression corresponding to Fig. 8 is w'.

Example Consider a function $f : B_4 \rightarrow B$ whose truth table is given below.

x	y	z	w	$f(x, y, z, w)$
0	0	0	0	1
0	0	0	1	1
0	0	1	0	1
0	0	1	1	1
0	1	0	0	0
0	1	0	1	0
0	1	1	0	0
0	1	1	1	0
1	0	0	0	1
1	0	0	1	1
1	0	1	0	0
1	0	1	1	0
1	1	0	0	0
1	1	0	1	0
1	1	1	0	0
1	1	1	1	1

Karnaugh map of f and a decomposition are given below.

1	1	1	1
0	0	0	0
0	0	1	0
1	1	0	0

The corresponding boolean expression is $(z' \wedge y') \vee (x' \wedge y' \wedge z) \vee (x \wedge y \wedge z \wedge w)$.

Solved problems

1. Construct Karnaugh maps for the functions whose truth tables are given below. Use Karnaugh map method to find a boolean expression for the function f using an appropriate decomposition.

(i)

x	y	$f(x, y)$
0	0	1
0	1	0
1	0	0
1	1	1

Karnaugh map of f is as follows.

	y'	y
x'	1	0
x	0	1

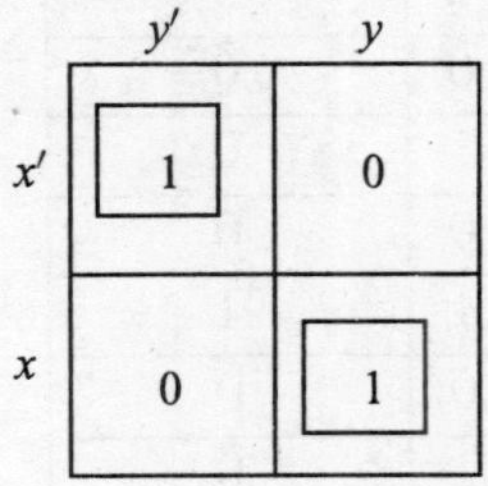

Corresponding to the above decomposition, the boolean expression is $(x' \wedge y') \vee (x \wedge y)$.

(ii)

x	y	z	$f(x, y, z)$
0	0	0	1
0	0	1	1
0	1	0	0
0	1	1	0
1	0	0	1
1	0	1	0
1	1	0	1
1	1	1	0

The Karnaugh map of f is as follows.

1	1	0	0
1	0	0	1

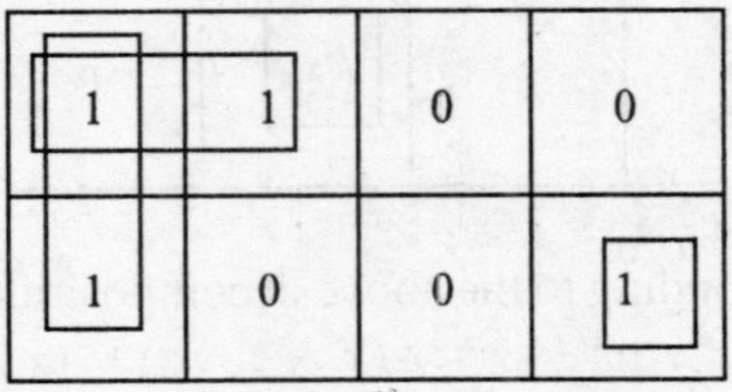

Corresponding to the above decomposition, the boolean expression is $(x' \wedge y') \vee (y' \wedge z') \vee (x \wedge y \wedge z')$.

(iii)

x	y	z	w	$f(x, y, z, w)$
0	0	0	0	0
0	0	0	1	0
0	0	1	0	1
0	0	1	1	0
0	1	0	0	0
0	1	0	1	0
0	1	1	0	1
0	1	1	1	0
1	0	0	0	0
1	0	0	1	0
1	0	1	0	0
1	0	1	1	1
1	1	0	0	0
1	1	0	1	0
1	1	1	0	1
1	1	1	1	1

Karnaugh map of f is as follows.

0	0	0	1
0	0	0	1
0	0	1	1
0	0	1	0

Corresponding to the above decomposition, the boolean expression is $(x' \wedge w' \wedge z') \vee (x \wedge y \wedge z) \vee (x \wedge z \wedge w)$.

2. In the following problems, the Karnaugh maps of functions are given. A decomposition of the 1-values into rectangles is also shown. Write the boolean expression for these functions that arise from the maps and rectangular decompositions.

(i)

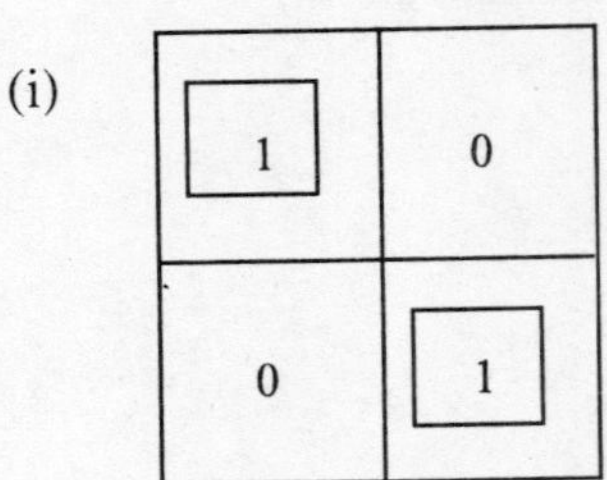

Answer: The boolean expression is $(x' \wedge y') \vee (x \wedge y)$.

(ii)

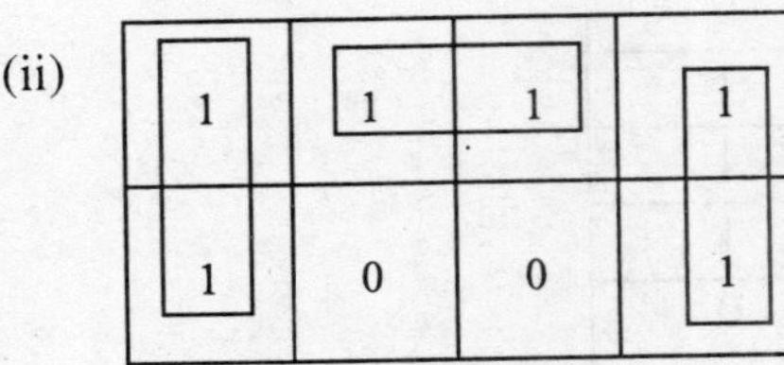

Answer: The boolean expression is

$(x' \wedge z) \vee (y' \wedge z') \vee (y \wedge z')$

$\equiv (x' \wedge z) \vee [z' \wedge (y' \vee y)]$

$\equiv (x' \wedge z) \vee z'$.

(iii)

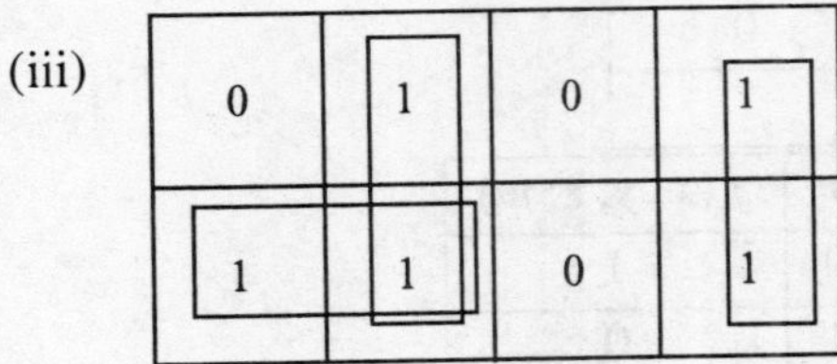

Answer: The boolean expression is $(x \wedge y') \vee (y' \wedge z) \vee (y \wedge z')$.

(iv)

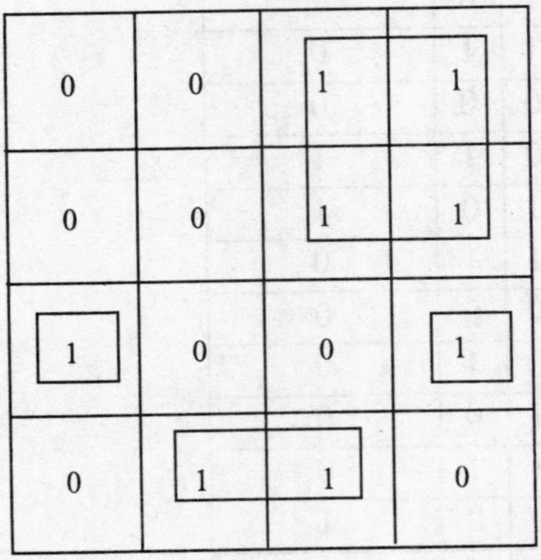

0	0	1	1
0	0	1	1
1	0	0	1
0	1	1	0

Answer: The boolean expression is

$(x' \wedge z) \vee (y' \wedge w \wedge x) \vee$
$[(x \wedge y \wedge z' \wedge w')$
$\vee (x \wedge y \wedge z \wedge w')])$

$\equiv (x' \wedge z) \vee (y' \wedge w \wedge x) \vee [(x \wedge y \wedge w') \wedge (z' \vee z)]$

$\equiv (x' \wedge z) \vee (y' \wedge w \wedge x) \vee (x \wedge y \wedge w')$.

Exercises

1. Construct Karnaugh maps for the functions whose truth tables are given below. Use Karnaugh map method to find a boolean expression for the function f using an appropriate decomposition.

(i)

x	y	$f(x, y,)$
0	0	1
0	1	0
1	0	1
1	1	0

(ii)

x	y	z	$f(x, y, z)$
0	0	0	0
0	0	1	1
0	1	0	1
0	1	1	1
1	0	0	0
1	0	1	0
1	1	0	0
1	1	1	1

(iii)

x	y	z	w	$f(x, y, z, w)$
0	0	0	0	1
0	0	0	1	0
0	0	1	0	1
0	0	1	1	0
0	1	0	0	0
0	1	0	1	1
0	1	1	0	1
0	1	1	1	0
1	0	0	0	0
1	0	0	1	0
1	0	1	0	0
1	0	1	1	0
1	1	0	0	1
1	1	0	1	0
1	1	1	0	1
1	1	1	1	0

2. In the following problems, Karnaugh maps of some functions are given. A decomposition of 1-values into rectangles is also shown. Write the boolean expression for these functions which arise from the maps and the rectangular decompositions.

(i)

1	1
1	0

(ii)

1	1	0	1
0	1	0	1

(iii)

0	1	0	1
1	1	0	1

(iv)

0	0	1	1
0	0	1	1
1	0	0	1
0	1	1	0

9 Elements of Graph Theory

Graph theory is an interesting branch of mathematics with potential for application in varied fields. Consider the electrical wiring in a house. Techniques of graph theory are used in designing the layout of the electrical circuits. Consider the following problems of day-to-day interest. Later on we will see how graph theory can be used in solving these problems.

1. There are two islands C and D. A and B are two banks. There are two bridges which can be used to traverse between bank A and island C and two bridges which can be used to traverse between bank B and island C. Again there is one bridge between bank A and island D and one bridge between bank B and island D. Further, assume that there is one bridge which connects islands C and D. Thus there are totally seven bridges and the problem is to determine whether it is possible for a person to start from either bank A or from bank B or from island C or from island D, pass through each of the seven bridges exactly once and return to the starting point. The situation mentioned in this problem can be depicted as in Fig. 9.1.

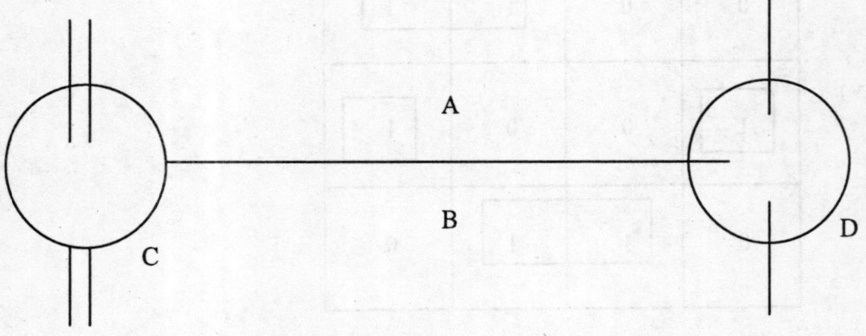

Fig. 9.1

   This problem is referred to as *Konigsberg bridge problem*.

2. A salesman starts from a city. He has to visit a number of cities and return to the starting city. The distance between any two cities is given.

Now the problem is how to visit all the cities and return to the starting city with minimum distance travelled. This problem is referred to as the Travelling Salesman Problem (TSP).

3. n People in an office sit at a round table for lunch everyday. They want to sit in such a way that each day, each person will have different neighbours. Certainly, this arrangement cannot last for ever. Now the problem is to determine the number of lunches for which this arrangement can last. This problem is referred to as the Seating Problem.

The above problems can also be solved using graph theory. Graph theory has developed so much that volumes can be written on the subject. Because of its very close association with Discrete Mathematical Structures, it has been introduced here at an elementary level. We give here the important concepts and results without going into their rigorous proofs.

All of us are very familiar with graphs from our childhood days. Is there anyone who has not heard about straight lines or triangles in school geometry? Straight lines and triangles are examples of graphs. What does a straight line do? It simply connects two points. What does a triangle do? There are three points and there is a straight line connecting any two points. The two points of a straight line are the vertices and the line connecting the two points an edge. Thus a straight line is a graph consisting of two vertices and one edge. A triangle is a graph consisting of three vertices and three edges. We thus see that every edge connects two vertices. These two vertices are called the end vertices of the edge.

Formally, a graph G is a pair (V, E) where V is a finite set whose elements are called vertices and E is also a finite set whose elements are called edges. If $e \in E$, then we can identify two vertices v_i and v_j which are the end vertices of e. Since an edge cannot exist without end vertices, we can associate with G a function $f : E \to V \times V$ which is defined as follows.

Consider an edge $e_i \in E$. If v_k and v_l are the end vertices of e_i, we define $f(e_i) = (v_k, v_l)$.

Note that in a graph $G = (V, E)$, it is possible that the end vertices of an edge are the same. In other words, there may be an edge e for which $f(e) = (v, v)$. Such an edge is called a self-loop or simply a loop. It is also not necessary that there must be at the most one edge between a pair of vertices. Thus, there can exist vertices v_i and v_j and edges e_k and e_l such that $f(e_k) = (v_i, v_j)$ and $f(e_l) = (v_i, v_j)$. This means vertices v_i and v_j are end vertices of both the edges e_k and e_l. The edges e_k and e_l are referred to as parallel edges. In other words, parallel edges have the same pair of end vertices.

Note that we can have any number of loops at every vertex and any number of parallel edges between any two vertices. Hence it will not be possible to say anything about the number of edges of a graph if we allow loops and parallel edges.

A graph $G = (V, E)$ is said to be a simple graph if it has no self-loops or parallel edges. Unless otherwise specified, all our graphs will be simple graphs.

Example Consider the figure given below which is a simple graph.

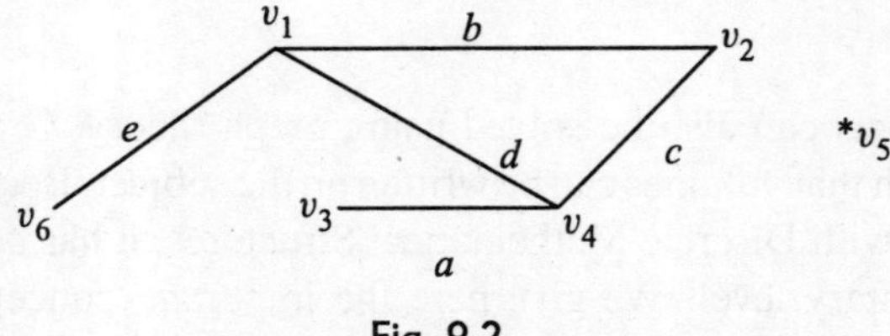

Fig. 9.2

Note that this is a graph where $V = \{v_1, v_2, v_3, v_4, v_5, v_6\}$ and $E = \{a, b, c, d, e\}$. We have

$$f(b) = (v_1, v_2), \quad f(c) = (v_2, v_4), \quad f(a) = (v_3, v_4),$$

and so on.

We note that v_1 is an end vertex of b. We also say that the edge b is incident on v_1. Since v_2 is also an end vertex of b, the edge b is also incident on v_2. Now how many edges are incident on v_1? This is the same as the number of edges for which v_1 is an end vertex. This number is seen to be 3. We say that the degree of v_1 is 3 and denote it by writing deg $(v_1) = 3$. A formal definition of the degree of a vertex is as follows.

Definition The degree of a vertex v denoted by $\deg(v)$ is the number of edges for which v is an end vertex. In other words, $\deg(v)$ is the number of edges which are incident on v.

By convention, a loop contributes to two degrees.

For the above graph (Fig. 9.2), we have $\deg(v_2) = 2$, $\deg(v_3) = 1$, $\deg(v_4) = 3$, $\deg(v_5) = 0$ and $\deg(v_6) = 1$. Note that there is no edge for which v_5 is an end vertex. We can also see that v_5 is an isolated vertex in the sense that it is not possible to reach vertex v_5. Thus an isolated vertex is simply a vertex whose degree is 0. There is only one edge for which v_3 is an end vertex and hence $\deg(v_3) = 1$. A vertex whose degree is 1 is called a pendant vertex. Thus v_3 is a pendant vertex. v_6 is another pendant vertex.

We have

$$\deg(v_1) + \deg(v_2) + \deg(v_3) + \deg(v_4) + \deg(v_5) + \deg(v_6)$$
$$= 3 + 2 + 1 + 3 + 0 + 1 = 10,$$

which is nothing but twice the number of edges. Note that the graph has 5 edges. Now consider the graph shown in Fig. 9.3 which is not a simple graph.

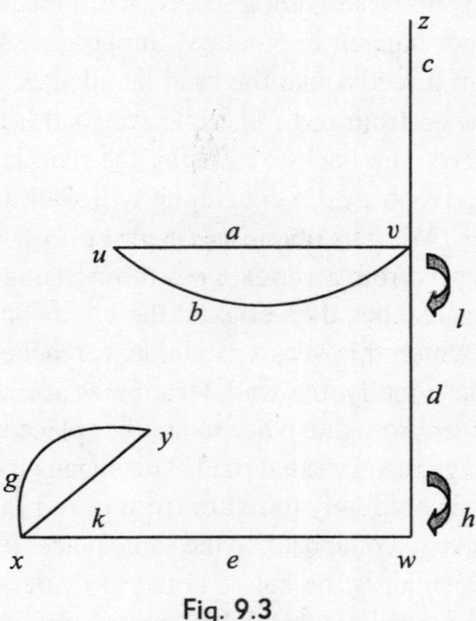

Fig. 9.3

This graph has 6 vertices which are labelled as u, v, w, x, y, and z and 9 edges which are labelled a, b, c, d, e, g, h, k and l. Note that h and l are loops, a and b are parallel edges and so are edges g and k. $\deg(u) = 2$, $\deg(v) = 6$ (there is a loop at v which contributes to 2 degrees), $\deg(w) = 4$, $\deg(x) = 3$, $\deg(y) = 2$ and $\deg(z) = 1$. We have

$$\deg(u) + \deg(v) + \deg(w) + \deg(x) + \deg(y) + \deg(z)$$
$$= 2 + 6 + 4 + 3 + 2 + 1 = 18;$$

which is once again equal to twice the number of edges.

We can prove this for any graph.

Theorem Let $G = (V, E)$ be a graph. Then the sum of the degrees of all vertices in G is twice the number of edges.

Proof The proof is very simple. Assume that G has m edges. Consider any edge e_k. We know that an edge cannot exist without end vertices. Let v_i and v_j be the end vertices of e_k. Because of the edge e_k, what has happened? The degree of v_i has increased by 1 and so has the degree of v_j increased by 1. Thus one edge e_k contributes to 2 degrees. m edges will contribute to $2m$ degrees.

In the graph given in Fig 9.2, assume that v_1, v_2, v_3, v_4, v_5 and v_6 denote the names of places in a city. Assume that they are all at walkable distances from one another. Assume that a, b, c, d and e denote the names of the roads connecting the places. Thus road d connects places v_1 and v_4; road c connects places v_2 and v_4 and so on. There is no road leading to v_5, which means v_5 is not accessible. Of course, this will not happen in practice! Suppose a salesman wants to walk from v_4 to v_3. He can directly take the road labelled a. We thus obtain v_4av_3. He can also decide to go from v_4 to v_1 using the road labelled d, proceed to v_2 using the road labelled b, go back to v_4 using the road labelled c and then take the road labelled a to reach v_3. Of course, he will walk this way only if he has some work in v_1 or v_2. We thus obtain $v_4dv_1bv_2cv_4av_3$. These are examples of walks. Thus a walk starts from a vertex, goes through one of the edges on which the vertex is incident, reaches the vertex at the end of that edge, goes through one of the edges on which this vertex is incident, reaches the other end-vertex of this edge and so on. Finally, the walk terminates at a vertex. While walking, nobody would like to go from one place to another place using a road and return to the same place using the very same road. This means if one goes from v_1 to v_2 using the edge b, one is not likely to return from v_2 to v_1 using the same edge b. However, one may have to come back to the same place. In the above graph, note that wherever the salesman is, he has to come to v_4 in order to go to v_3. Thus vertex v_4 is repeated in the above walk. In other words, in a walk a vertex can be repeated but not an edge.

Consider a graph whose vertices are the names of cities. Suppose a salesman wants to go from city c_i to city c_j. He might be advised to go from city c_i to some other city, from there to yet another city and so on, and finally from the last city to city c_j. Nobody will advise him to come back to a city already visited by him or to take the same road through which he has already travelled. Thus he has been shown a path to go from city c_i to city c_j. A path is simply a walk in which no vertex can appear more than once. $v_6ev_1dv_4av_3$ is a path to go from v_6 to v_3 in the graph given in Fig 9.2. $v_6ev_1bv_2cv_4av_3$ is another path to go from v_6 to v_3. Since there are two paths to go from v_6 to v_3, which path will one choose? Suppose a, b, c, d and e denote the distances between the places. For example, let a denote the distance between the places v_4 and v_3. Then one would naturally like to take that path whose length is the minimum. The length of a path is nothing but the sum of the lengths of the various edges that appear on the path. If $a = 5, b = 3, c = 6, d = 2$ and $e = 7$, then the length of the path $v_6ev_1dv_4av_3$ is $7 + 2 + 5 = 14$, whereas the length of the path $v_6ev_1bv_2cv_4av_3$ is $7 + 3 + 6 + 5 = 21$. Hence one would like to take the path $v_6ev_1dv_4av_3$ to go from v_6 to v_3.

A graph in which a real number is associated with every edge is referred to as a weighted graph.

If a walk starts from a vertex and ends at the same vertex, it is called a closed walk. If the starting and ending vertices are not the same, it is called an open walk. For example, $v_4dv_1bv_2cv_4av_3$ is an open walk. In a path, if the initial and final vertices are the same, then the path is called a circuit. For example, $v_4dv_1bv_2cv_4$ is a circuit. Thus a circuit is a closed walk in which no vertex (except of course the initial and final vertices) can appear more than once. We can see that if a graph has a circuit, then there will exist at least one pair of vertices between which there will be more than one path. The converse is also true. This means that if there are two vertices in a graph between which there exist more than one path, then the graph will have a circuit.

Note that even though there is no direct edge connecting v_6 and v_2 in the graph of Fig. 9.2, one can still go from v_6 to v_2 using the path $v_6ev_1bv_2$. Except for v_5, we can go from any vertex to any other vertex by a path. If we discard v_5, then the graph becomes a connected graph. Thus a connected graph is a graph in which there is a path from any vertex to any other vertex. If a graph is not connected, it is called a disconnected graph. Thus in a disconnected graph we can find at least two vertices between which there is no path. The above graph (including v_5) is a disconnected graph.

As another example of a disconnected graph, consider a graph whose vertices are the names of cities in India and Malaysia. Thus the vertices of the graph are either the names of cities in India or Malaysia. An edge between two vertices is nothing but a road which connects the two cities represented by the vertices. We assume that any two cities in India are connected by road and that any two cities in Malaysia are also connected by road. We also assume that there is no road which connects a city in India to a city in Malaysia. Denote the cities in India together with the roads which connect any two cities in India by I. Similarly, M denotes the cities in Malaysia together with the roads which connect any two cities in Malaysia. I is a connected graph, and so is M. I and M together constitute a disconnected graph.

Consider a graph $G = (V, E)$. Let V_1 be a subset of the set of vertices V. Let E_1 denote the set of edges of E which are incident on the vertices which are in V_1. Then $G_1 = (V_1, E_1)$ is called a sub-graph of $G = (V, E)$. Some of the sub-graphs for the graph given in Fig. 9.2 are shown in Fig. 9.4.

Consider a graph $G = (V, E)$. A maximal connected sub-graph of G is a connected sub-graph G_1 of G which has the property that it is not possible to obtain a larger connected sub-graph of G by adding more edges to G_1. In other words, even if we try to add one more edge to G_1, the resulting sub-graph will no longer remain connected. Such a maximal connected sub-graph is called

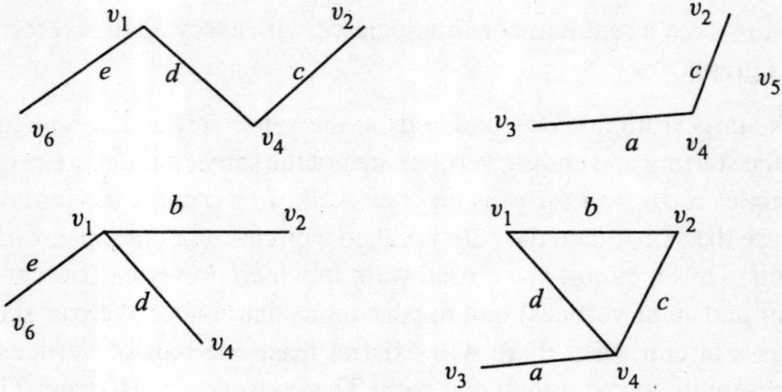

Fig. 9.4

a component. For the above disconnected graph, I and M are components. We note that there are no vertices which are common to I and M, no edges which are common to I and M, and furthermore, I and M together give back the original graph. We say that the disconnected graph has been partitioned into its components I and M. We can see that any disconnected graph can be partitioned into the components $c_1, c_2, \ldots, c_n$. Note that each of $c_1, c_2, \ldots, c_n$ is a connected graph.

In the graph given in Fig. 9.2, there are pairs of vertices (for example, v_2 and v_6, v_1 and v_3, etc.) between which there are no edges. Hence this graph is not a complete graph. Look at the graph I mentioned above. Since we assume that there is a road connecting any two cities in India, I is a complete graph. Similarly, M is a complete graph.

Definition A graph $G = (V, E)$ is said to be a complete graph if there is an edge between any two vertices.

Shown in Fig. 9.5 are the complete graphs on 4 vertices and 5 vertices.

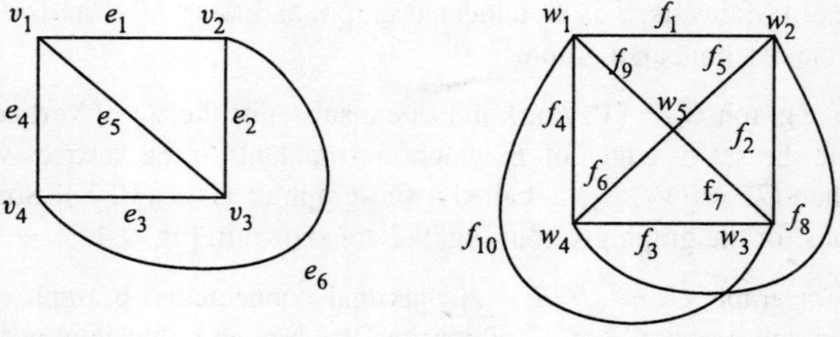

Fig. 9.5

Consider a complete graph on n vertices, say $G = (V, E)$. Thus $|V| = n$. Let $|E| = m$. Consider any vertex $v \in V$. Since G is a complete graph, there is an edge from v to each of the remaining $n-1$ vertices. Hence the degree of v is $n-1$. Note that v is an arbitrary vertex. Thus the sum of the degrees of all n vertices is $n-1+n-1+\cdots+n-1$(n times), which is nothing but $n(n-1)$. But we know this sum is $2m$. Thus $2m = n(n-1)$, proving that $m = \frac{n(n-1)}{2}$. We have thus proved the following.

Result The number of edges in a complete graph of n vertices is $\frac{n(n-1)}{2}$.

From the above result we can see that the number of edges in any simple graph consisting of n vertices is less than or equal to $\frac{n(n-1)}{2}$.

We will now consider the Konigsberg bridge problem stated earlier. Before considering a solution to this problem, we will introduce a terminology.

Consider a graph $G = (V, E)$. An Euler line is a closed walk (a walk which starts at a vertex and ends at the same vertex) which passes through every edge of the graph exactly once. A graph $G = (V, E)$ is said to be an Euler graph if it has an Euler line.

Consider the graph given in Fig. 9.6. We can see that the graph is an Euler graph. *uavcweyfxdwbu* is an Euler line for this graph.

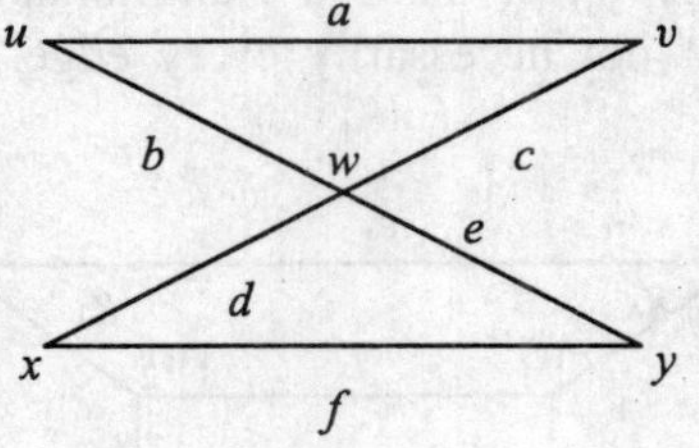

Fig. 9.6

Now consider the Konigsberg bridge problem. We will represent the banks A, B and the islands C, D by vertices and each bridge by an edge. Then we obtain the graph shown in Fig. 9.7 which has 4 vertices and 7 edges.

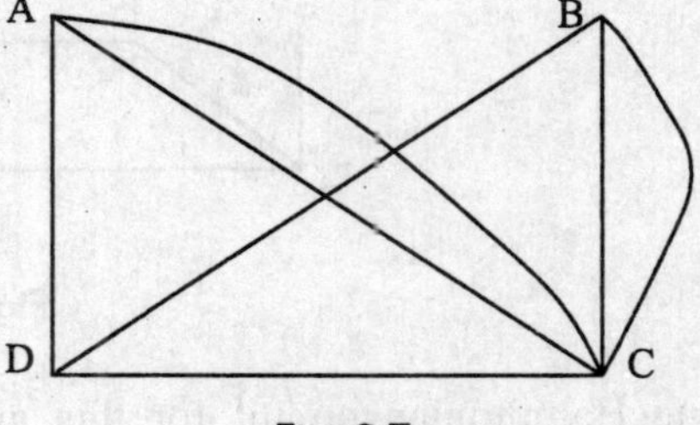

Fig. 9.7

Note that the problem of determining whether it is possible to start at a bank or an island, walk through each bridge exactly once and return to the starting point is the same as the problem of determining whether the graph is an Euler graph. This problem is also equivalent to the problem of determining whether it is possible to draw a graph starting at any vertex, drawing an edge only once (not going through an edge which is already drawn) and returning to the starting vertex all without lifting the hand. The graph in Fig. 9.7 is not an Euler graph can be seen from the following theorem stated without proof.

Theorem A connected graph $G = (V, E)$ is an Euler graph if and only if all its vertices have even degrees.

None of the vertices of the above graph has an even degree.

$$\deg(A) = \deg(B) = \deg(D) = 3 \text{ and } \deg(C) = 5.$$

Now consider an Euler graph $G = (V, E)$. G contains an Euler line. Consider any vertex v_i of G. Since G is a connected graph, there will be at least one edge, say e_j, which is incident on v_i. Since the Euler line contains every edge of the graph, it contains the edge e_j. Hence it should also contain the end vertex of e_j, namely v_i. (Can a walk contain an edge without its end vertices?) This means that an Euler line contains every vertex of the graph.

Associated with the concept of Euler graph is the concept of Hamiltonian circuit. A Hamiltonian circuit is a circuit which traverses through every vertex of the graph exactly once. Thus a Hamiltonian circuit contains every vertex of the graph but not necessarily every edge. Consider the graph in Fig. 9.8.

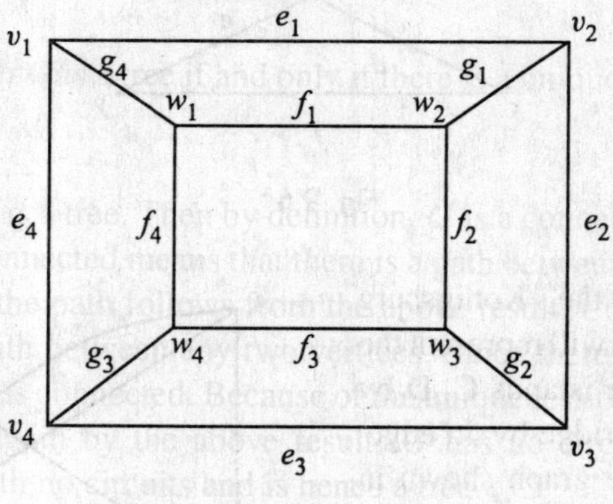

Fig. 9.8

One Hamiltonian circuit for this graph is $v_1 e_1 v_2 g_1 w_2 f_2 w_3 g_2 v_3 e_3 v_4 g_3 w_4 f_4 w_1 g_4 v_1$.

What about the existence of Hamiltonian circuits in a graph? The following theorem asserts that every complete graph with at least three vertices has a Hamiltonian circuit. In fact, the theorem counts the number of Hamiltonian circuits such that no two circuits have any edge in common.

Theorem A complete graph with n vertices has $\frac{n(n-1)}{2}$ edge disjoint Hamiltonian circuits.

Consider a weighted graph $G = (V, E)$ which has a Hamiltonian circuit. The weight of a Hamiltonian circuit is defined as the sum of weights of all the edges in the given Hamiltonian circuit. Now consider the travelling salesman problem. If vertices denote cities and edges denote roads connecting the cities, then we obtain a weighted graph in which the weight associated with any edge is nothing but the distance between cities which are connected by the edge. If v_i and v_j are vertices of the graph (note that v_i and v_j are the cities) and e_k is an edge connecting these two vertices (note that e_k is a road which connects the cities v_i and v_j), then the weight associated with e_k is nothing but the distance between the cities v_i and v_j through this road. We find all the Hamiltonian circuits in this weighted graph and determine the weight of each circuit. Take the Hamiltonian circuit whose weight is the smallest. This circuit will give the route that the travelling salesman will have to follow so as to visit all the cities while keeping the distance travelled to a minimum.

Since a Hamiltonian circuit in a graph with n vertices has to contain all the n vertices, its length (number of edges) has to be n. If we remove any edge from a Hamiltonian circuit, we obtain a Hamiltonian path of length $n - 1$. Note that a Hamiltonian path also contains all vertices of the graph but it starts at one vertex and terminates at another vertex.

Now consider the graph given in Fig. 9.9.

We can see that there is a path from any vertex to any other vertex in this graph and hence this graph is a connected graph. We can also see that this graph has a circuit and there are two vertices v_1 and v_3 between which there are two paths. One path is $v_1e_1v_2e_2v_3$ and the other path is $v_1e_4v_5e_3v_3$.

Now consider the graph shown in Fig. 9.10.

This graph is also connected and we can see that there is a unique path between any two vertices. One path between the vertices w_1 and w_5 is

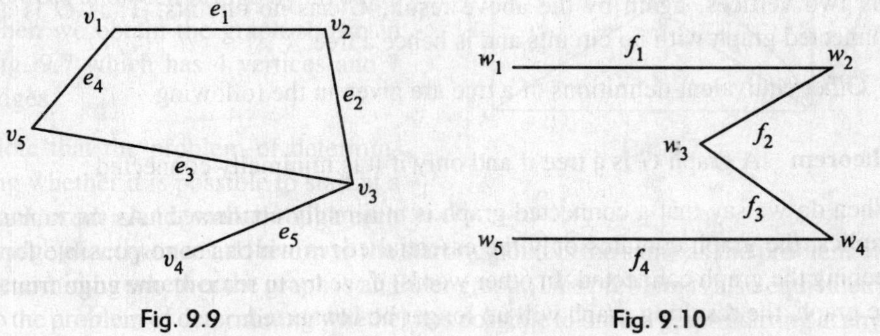

Fig. 9.9 Fig. 9.10

$w_1 f_1 w_2 f_2 w_3 f_3 w_4 f_4 w_5$. We can see that there is no other path between w_1 and w_5. Further, this graph does not have any circuits.

The above two examples make us wonder whether there is any relationship between the existence of circuits and the number of paths (between two vertices) in a connected graph. There does exist a relationship as can be seen from the following.

Result A connected graph $G = (V, E)$ has a circuit if and only if there are vertices in G between which there exists more than one path.

Proof Suppose there is a circuit which starts at a vertex, say a, and terminates at the same vertex a. Then we can find a vertex b in the circuit such that there is a path, say p_1 from a to b and a path p_2 from b to a, which is also a path from a to b. Thus there are two paths p_1 and p_2 between vertices a and b.

Conversely, assume that there are two vertices, say x and y, between which there exist two paths, say p_3 and p_4. Thus using the path p_3 we can go from vertex x to vertex y and using the path p_4 we can go from vertex y to vertex x. Combining the two paths p_3 and p_4, we obtain a circuit which starts at vertex x and terminates at the same vertex.

We are now in a position to define a tree.

Definition A tree is a connected graph which has no circuits.

Theorem A graph G is a tree if and only if there is a unique path between any two vertices.

Proof Suppose G is a tree. Then by definition, G is a connected graph with no circuits. G being connected means that there is a path between any two vertices. The uniqueness of the path follows from the above result. Conversely, suppose there is a unique path between any two vertices. Since there is a path between any two vertices, G is connected. Because of the uniqueness of the path between any two vertices, again by the above result, G has no circuits. Thus G is a connected graph with no circuits and is hence a tree.

Other equivalent definitions of a tree are given in the following.

Theorem A graph G is a tree if and only if it is minimally connected.

When do we say that a connected graph is minimally connected? As the name implies, the graph contains only the essential edges which are responsible for keeping the graph connected. In other words, if we try to remove one edge from the graph, the resulting graph will no longer be connected.

Consider a connected graph G with n vertices. We can see that n edges will always form a circuit. Since a tree has no circuits, it follows that a tree with n vertices cannot have more than $n - 1$ edges. The actual number of edges in a tree with n vertices is given by the following.

Theorem A tree with n vertices has $n - 1$ edges.

The converse of the above theorem is also true. In other words, we can prove the following.

Theorem A connected graph with n vertices and $n - 1$ edges is a tree.

If we have a graph with n vertices, $n - 1$ edges and no circuits, then we can prove that the graph is connected. In other words, the following holds good.

Theorem A graph with n vertices, $n - 1$ edges and no circuits is a tree.

Consider a connected graph G which is not a tree. Hence G has circuits. Many a time we will be interested only in visiting all the vertices of G but not all the edges of G. This is particularly true if the vertices of G denote cities and the edges denote roads connecting the cities. Suppose a person wants to start from one city, visit all other cities and end up in the last city. Note that he does not want to return to the starting city. In this case, why should he pass through all possible roads? He also would not like to come back to a city that he has already visited. What we are asking for is a spanning tree whose definition is as follows.

Definition A spanning tree for a graph G is a sub-graph of G which is a tree and which contains all the vertices of G.

Consider a connected graph G of n vertices. If G has no circuits, then G is a tree. Since G contains all vertices, it is also a spanning tree of G. If G contains a circuit, remove an edge from the circuit. Note that by removing an edge which is part of a circuit, the connectedness is not affected. Thus the resulting sub-graph is still connected and contains all the vertices of G. If it has no circuits, then it is a spanning tree of G. If the sub-graph does have a circuit, then remove an edge from the circuit. The resulting sub-graph is still connected and contains all the vertices of G. If it has no circuits, it is a spanning tree of G. Otherwise, repeat the above process. We will finally end up with a sub-graph of G which is connected, which has no circuits and which contains all the vertices of G and is hence a spanning tree of G.

We have thus proved the following.

Theorem Every connected graph G has at least one spanning tree.

Consider a connected graph $G = (V, E)$ and consider any spanning tree T of G. Note that if we try to add one more edge of G (which is not in T) to T, we will obtain a circuit. However, after adding such an edge we can remove an edge which was already in T. By doing this, we will obtain another spanning tree of G. We thus obtain several spanning trees of G. If we assume that G is a weighted graph, we can compute the weight of each of these spanning trees. Note that the weight of a spanning tree is nothing but the sum of weights of all the edges in the spanning tree. Then we can select that spanning tree whose weight is the minimum. This spanning tree will give the route the salesman has to take if he decides to start from one city, visit all other cities and end up in a city other than the one he started from, while covering the minimum distance. Consider the graph given in Fig. 9.11.

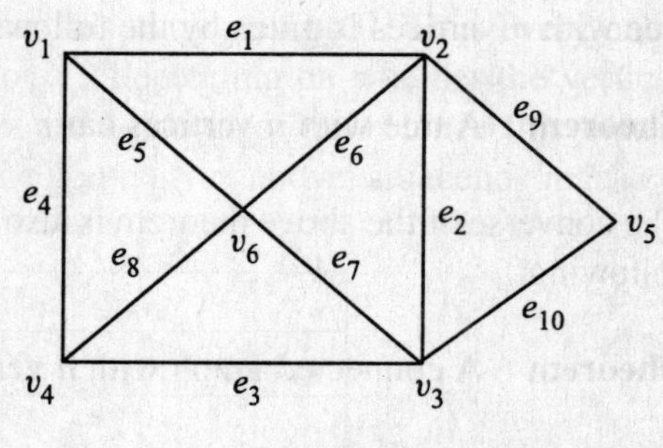

Fig. 9.11

Three different spanning trees for this graph are given in Figs. 9.12–9.14.

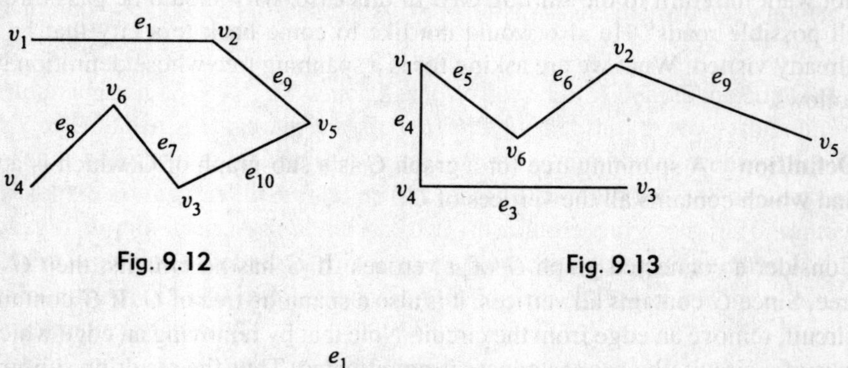

Fig. 9.12

Fig. 9.13

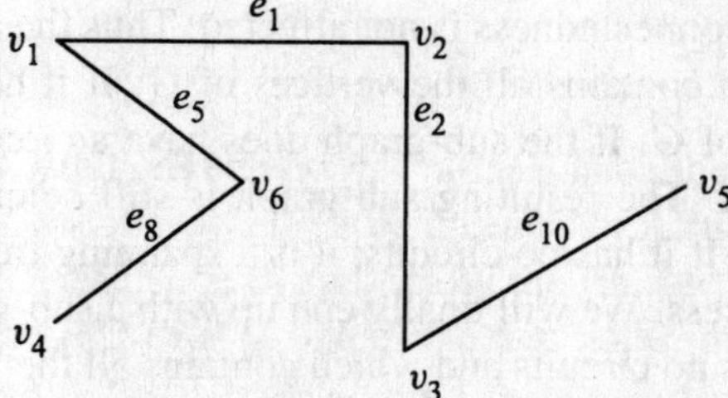

Fig. 9.14

9.1 Representation of a Graph on Computer

We will now consider the problem of representing the data pertaining to a graph on a computer. There are several matrices which are associated with a graph. We will study them one by one.

Incidence matrix representation of a graph

Consider a graph $G = (V, E)$ which has m vertices and n edges. We will associate with G a matrix which has m rows and n columns and which is defined as follows.

Every row of the matrix corresponds to a vertex and every column of the matrix corresponds to an edge. Consider the entry in the matrix which corresponds to vertex v_i and edge e_j. This entry will be 1 if and only if vertex v_i is an end vertex of edge e_j. Otherwise, this entry will be 0. Thus every entry of the incidence matrix is either a 1 or a 0 depending on whether the vertex corresponding to the row is incident on the edge corresponding to the column or not. Hence this matrix gets the name of incidence matrix. For the graph given in Fig. 9.11, the incidence matrix is as follows.

	e_1	e_2	e_3	e_4	e_5	e_6	e_7	e_8	e_9	e_{10}
v_1	1	0	0	1	1	0	0	0	0	0
v_2	1	1	0	0	0	1	0	0	1	0
v_3	0	1	1	0	0	0	1	0	0	1
v_4	0	0	1	1	0	0	0	1	0	0
v_5	0	0	0	0	0	0	0	0	1	1
v_6	0	0	0	0	1	1	1	1	0	0

How many 1s are there in the row corresponding to v_4? Three 1s. What is the degree of v_4? Also three. The number of 1s in any row of an incidence matrix is the degree of the vertex corresponding to that row.

Adjacency matrix representation of a graph

Again consider a graph $G = (V, E)$ which has m vertices. We will associate with G a matrix which has m rows and m columns defined as follows.

Every row of the matrix corresponds to a vertex and every column of the matrix also corresponds to a vertex. Consider the entry in the matrix which corresponds to vertices v_i and v_j. This entry will be 1 if and only if there is an edge which connects vertices v_i and v_j. Otherwise, this entry will be 0. Thus every entry of the adjacency matrix is either a 1 or a 0 depending on whether the vertices corresponding to the row and column are adjacent or not. Hence this matrix gets the name adjacency matrix. For the graph given above, adjacency matrix is as follows.

	v_1	v_2	v_3	v_4	v_5	v_6
v_1	0	1	0	1	0	1
v_2	1	0	1	0	1	1
v_3	0	1	0	1	1	1
v_4	1	0	1	0	0	1
v_5	0	1	1	0	0	0
v_6	1	1	1	1	0	0

Since the graph has no loops, the main diagonal entries are all 0s. Since the graph is a undirected one (edges do not have any directions), the adjacency matrix is a symmetric matrix. Note also that the adjacency matrix of a graph is a square matrix unlike the incidence matrix.

We will now see how an adjacency matrix will help us in determining whether a graph is connected or not.

Consider a graph $G = (V, E)$ which has n vertices. Let A denote the adjacency matrix of G. Then A is an $n \times n$ matrix. We know that the ith row, jth column entry of A being 0 implies that the vertices v_i and v_j are not adjacent, which is equivalent to saying that there is no path of length 1 from v_i to v_j. A^2 denotes the product of A with itself. Again, the ith row, jth column entry of A^2 is 0 implies that there is no path of length 2 from v_i to v_j. Now consider A^3. The ith row, jth column entry of A^3 is 0 implies that there is no path of length 3 from v_i to v_j. Now compute the matrix

$$X = A + A^2 + A^3 + \cdots + A^{n-1}.$$

Suppose the ith row, jth column entry of X is 0. Then the ith row, jth column entry of A must be 0, the ith row, jth column entry of A^2 must be 0, the ith row, jth column entry of A^3 must be 0, and finally the ith row, jth column entry of A^{n-1} must be 0. Note that even if one of them is not 0, the sum will not be 0. This means that there is no path of length 1 from v_i to v_j, there is no path of length 2 from v_i to v_j, there is no path of length 3 from v_i to v_j, and finally there is no path of length $n - 1$ from v_i to v_j. Thus there cannot be any path from v_i

to v_j. Note that if at all there is a path from v_i to v_j, its length must be less than or equal to $n - 1$. This means that the graph is disconnected. Thus adjacency matrix is very useful in determining whether a graph is connected or not. The graph $G = (V, E)$ is connected if and only if no entry of the matrix X is 0.

Path matrix representation of a graph

Unlike incidence matrix and adjacency matrix, path matrix is normally defined for two vertices u and v. Consider two vertices u and v and consider all the possible paths between u and v. Note that since a graph which is not a tree has at least one circuit, there can exist more than one path between u and v. Let these paths be $P_1, P_2, \ldots, P_k$. If n denotes the number of edges of G, then the path matrix between u and v denoted by $P(u, v)$ is a $k \times n$ matrix whose rows correspond to the various paths and whose columns correspond to the various edges. This matrix is defined as follows.

The ith row, jth column entry of $P(u, v)$ is 1 if and only if the edge e_j is in path P_i. Otherwise, this entry is 0. Thus every entry of the path matrix is also 1 or 0 depending on whether an edge belongs to a path or not. For the graph given in Fig. 9.11, consider the vertices v_1 and v_3. Various paths between these two vertices are the following.

$P_1 = v_1e_1v_2e_2v_3$
$P_2 = v_1e_1v_2e_9v_5e_{10}v_3$
$P_3 = v_1e_5v_6e_7v_3$
$P_4 = v_1e_5v_6e_6v_2e_2v_3$
$P_5 = v_1e_5v_6e_6v_2e_9v_5e_{10}v_3$
$P_6 = v_1e_4v_4e_3v_3$
$P_7 = v_1e_4v_4e_8v_6e_7v_3$
$P_8 = v_1e_4v_4e_8v_6e_6v_2e_2v_3$
$P_9 = v_1e_4v_4e_8v_6e_6v_2e_9v_5e_{10}v_3$
$P_{10} = v_1e_5v_6e_8v_4e_3v_3$

The path matrix is as follows.

	e_1	e_2	e_3	e_4	e_5	e_6	e_7	e_8	e_9	e_{10}
P_1	1	1	0	0	0	0	0	0	0	0
P_2	1	0	0	0	0	0	0	0	1	1
P_3	0	0	0	0	1	0	1	0	0	0
P_4	0	1	0	0	1	1	0	0	0	0
P_5	0	0	0	0	1	1	0	0	1	1
P_6	0	0	1	1	0	0	0	0	0	0
P_7	0	0	0	1	0	0	1	1	0	0
P_8	0	1	0	1	0	1	0	1	0	0
P_9	0	0	0	1	0	1	0	1	1	1
P_{10}	0	0	1	0	1	0	0	1	0	0

How many 1s are there in the row corresponding to P_5? Four. This number indicates the fact that the path P_5 contains four edges. We also know what these four edges are. They are precisely the edges in whose columns there are 1s.

Exercises

1. A graph which has 8 edges has 4 vertices each of degree 3. Each of the remaining vertices has degree 2. Determine the number of vertices in the graph.
2. Prove that a graph G is disconnected if and only if its vertex set V can be partitioned into two non-empty disjoint subsets V_1 and V_2 such that there is no edge in G whose one end-vertex is in V_1 and the other end-vertex is in V_2.
3. If a graph has exactly two vertices of odd degree, prove that there is a path which joins these two vertices.

10 Algebraic Structures

Consider a set S, for example the set N of all natural numbers. This set gains additional importance since it is possible to add two natural numbers, multiply two natural numbers and obtain new natural numbers. Now consider the set Z of all integers. We can add two integers, subtract one integer from another integer, etc. and obtain new integers. In the same way we will try to impose additional structures on a given set.

Consider the operation of addition (+) on the set of all natural numbers. We know that if m and n are two natural numbers, $m + n$ is also a natural number. '+' can be thought of as a function from $N \times N$ to N, i.e., $+ : N \times N \to N$. Any element of $N \times N$ is of the form (m, n), where $m, n \in N$ and $+(m, n)$ which is an element of N is denoted by $m + n$. We say + is a binary operation on the set of all natural numbers. Similarly − is a binary operation on the set Z of all integers. Again '−' can be thought of as a function from $Z \times Z$ to Z. $-(p, q)$ is denoted by $p - q$ where $p, q \in Z$.

Let S be any set. $*$ is said to be a *binary operation* on S if $*$ is a function from $S \times S$ to S. This means for any two elements $a, b \in S$, $*(a, b)$ is a unique element of S which is denoted by $a * b$.

Examples

1. Addition and multiplication are binary operations on the sets N, Z, Q and $\mathbb{R}$.
2. Subtraction is not a binary operation on N. For, if we subtract one natural number from another natural number, the result need not be a natural number. For example, $3 - 8 = -5$ is not a natural number. However, subtraction is a binary operation on Z, Q and $\mathbb{R}$.
3. Let S be any set and consider $A = P(S)$, the power set of S. If $S_1, S_2 \in A$, (S_1 and S_2 are subsets of S) then $S_1 \cup S_2 \in A$ ($S_1 \cup S_2$ is also a subset of S) and $S_1 \cap S_2 \in A$ ($S_1 \cap S_2$ is also a subset of S). Hence union and intersection are binary operations on $P(S)$.
4. Consider a set A and let $p(A)$ denote the set of all permutations on A (the set of all functions from A to A which are both one–one and onto). If

p_1 and p_2 are two permutations on A, then $p_1 \circ p_2$ is also a permutation on A. Hence $\circ$ is a binary operation on $p(A)$.

5. The operation of division (/) is not a binary operation on the set $\mathbb{R}$ of all real numbers. This is because for any real number a, $a/0$ is not a real number. However '/' is a binary operation on $\mathbb{R}^*$, the set of all non-zero real numbers.

Definition A non-empty set S together with a binary operation is called a *groupoid*. Thus a groupoid is little more than a set in the sense that it is a set together with a binary operation.

Examples

1. $(N, +)$, $(Z, +)$, $(Q, +)$ and $(\mathbb{R}, +)$ are all groupoids.
2. $(N, \times)$, $(Z, \times)$, $(Q, \times)$ and $(\mathbb{R}, \times)$ are all groupoids where $\times$ denotes multiplication operation.
3. $(N, -)$ is not a groupoid since '$-$' is not a binary operation on N.
4. $(Z, -)$, $(Q, -)$ and $(\mathbb{R}, -)$ are all groupoids.
5. For any set S, $(P(S), \cup)$ and $(P(S), \cap)$ are groupoids.
6. For any set A, $(p(A), \circ)$ is a groupoid.

Consider the set Z and the binary operations $+$ and $-$. We know that for any two integers m and n, $m + n = n + m$ but $m - n \neq n - m$. We say the binary operation $+$ is commutative whereas the binary operation $-$ is not commutative.

Definition Let $*$ be a binary operation on a set S. $*$ is said to be *commutative* if $a * b = b * a$ for all $a, b \in S$.

Again, consider the set Z and the binary operations $+$ and $-$. We know for any three integers m, n and p, $(m+n)+p = m+(n+p)$ but $(m-n)-p \neq m-(n-p)$. That is, whether we add m and n first and then add p or whether we add n and p first and then add m is immaterial to the result. We say the binary operation $+$ is associative but the binary operation $-$ is not associative.

Definitions

1. A binary operation $*$ on a set S is said to be *associative* if $a * (b * c) = (a * b) * c$ for all a, b and $c \in S$.
2. A *semigroup* is a non-empty set S together with a binary operation which is associative.

Thus a *semigroup* is little more than a groupoid in the sense that we have the additional restriction that the binary operation is associative.

Examples

1. $(N, +)$, $(Z, +)$, $(Q, +)$ and $(\mathbb{R}, +)$ are all commutative semigroups.
2. $(N, \times)$, $(Z, \times)$, $(Q, \times)$ and $(\mathbb{R}, \times)$ are all commutative semigroups.
3. $(N, -)$ is not a groupoid and hence cannot be a semigroup.
4. $(Z, -)$, $(Q, -)$ and $(\mathbb{R}, -)$ are groupoids which are not semigroups.
5. For any set S, $(P(S), \cup)$ and $(P(S), \cap)$ are commutative semigroups.
6. For any set A, $(p(A), \circ)$ is a semigroup which is not commutative.

Consider the semigroup $(Z, +)$. In this semigroup there is an element 0 which has the property that $n + 0 = 0 + n = n \ \forall\ n \in Z$. Now consider the semigroup $(Z, \times)$. In this semigroup, there is an element 1 which has the property that $n \times 1 = 1 \times n = n \ \forall\ n \in Z$. We say 0 is an identity element of $(Z, +)$ and 1 is an identity element of $(Z, \times)$. This leads to the following definition.

Definition In a semigroup $(S, *)$ an element e is called an *identity element* if $x * e = e * x = x \ \forall\ x \in S$. We can also define *left identity* as an element e which satisfies $e * x = x \ \forall\ x \in S$ and *right identity* as an element f which satisfies $x * f = x \ \forall\ x \in S$.

It is not necessary that a semigroup must have an identity element. However, a semigroup with an identity element is called *a monoid*. Thus a monoid is little more than a semigroup in the sense that the semigroup must have an identity element for it to become a monoid.

Examples

1. $(N, +)$ is a semigroup which is not a monoid. This is because 0 is the only element which has the property that $n + 0 = 0 + n = n \ \forall\ n \in N$ but $0 \notin N$.
2. $(Z, +)$, $(Q, +)$, $(\mathbb{R}, +)$ are all monoids in which 0 acts as identity element.
3. $(Z, \times)$, $(N, \times)$, $(\mathbb{R}, \times)$ are all monoids where $\times$ denotes multiplication operation. In each of these monoids, 1 acts as an identity element.
4. For any set S, $(P(S), \cup)$ is a monoid. For every $A \in P(S)$, $A \cup \phi = \phi \cup A = A$ and hence ϕ acts as an identity element.
5. For any set S, $(P(S), \cap)$ is a monoid. For every $A \in P(S)$, $A \cap S = S \cap A = A$, and hence S acts as identity element.
6. $(p(A), \circ)$ is a monoid in which I_A acts as identity element.

We can see that 0 is the only identity element of $(Z, +)$. No other element of Z has the property which 0 possesses. Similarly, 1 is the only identity element of $(Z, \times)$.

Result If a semigroup $(S, *)$ has an identity element, it is unique.

Proof Suppose that e_1 and e_2 are both identity elements of $(S, *)$. Since e_1 is an identity element, by definition we have

$$e_1 * e_2 = e_2 * e_1 = e_2.$$

Since e_2 is an identity element, by definition we have

$$e_1 * e_2 = e_2 * e_1 = e_1.$$

From the above equations it follows that $e_1 = e_2$.

In view of the above result, we can say that a monoid has a unique identity element which we will denote by e.

Now consider the monoid $(Z, +)$. We know that for every integer n, there corresponds another integer $-n$ which has the property $n + (-n) = (-n) + n = 0$, which is the identity element of $(Z, +)$. Again, consider the monoid $(Q^*, \times)$ where $Q^* = Q - \{0\}$, the set of all non-zero rational numbers. Any element of Q^* is of the form $\frac{p}{q}(p, q \in Z, p \neq 0, q \neq 0)$. Now $\frac{q}{p} \in Q^*$ and $\frac{p}{q} \times \frac{q}{p} = \frac{q}{p} \times \frac{p}{q} = 1$ which is the identity element of $(Q^*, \times)$. We say that $-n$ is an inverse of n in $(Z, +)$ and $\frac{q}{p}$ is an inverse of $\frac{p}{q}$ in $(Q^*, \times)$. This leads to the following definition.

Definition Let $(S, *)$ be a monoid. An element $y \in S$ is called an *inverse of* x if $x * y = y * x = e$, where e is the identity element of S. An element z is said to be a *left inverse* of x if $z * x = e$, and an element u is said to be a *right inverse* of x if $x * u = e$.

Note

1. In a monoid, only for the *identity element* e we require that $x * e = e * x = x \; \forall \, x$. This does not mean that the monoid should be commutative. For example, $(p(A), \circ)$ is a monoid which is not commutative.
2. In a monoid $(S, *)$, if y is an *inverse of* x, then we require that $x * y = y * x = e$. This does not mean that the monoid should be commutative.

It is not necessary that in a monoid every element must have an inverse. However, we can prove the following.

Result In a monoid $(S, *)$, if an element x has an inverse, it is unique.

Proof Let $(S, *)$ be a monoid and let $x \in S$. Suppose x has two inverses y_1 and y_2. We will prove that $y_1 = y_2$. Since y_1 is an inverse of x, by definition of an inverse we have,

$$x * y_1 = y_1 * x = e.$$

Since y_2 is an inverse of x, we have

$$x * y_2 = y_2 * x = e.$$

Now $y_1 = y_1 * e = y_1 * (x * y_2) = (y_1 * x) * y_2 = e * y_2 = y_2$.

From the above result we see that if an element x in a monoid $(S\ *)$ has an inverse, then it is unique. This unique inverse of x depends on x and is denoted by x^{-1}.

Definition A group is a monoid $(G, *)$ in which every element has an inverse.

Note To prove that $(G, *)$ is a group, we have to prove the following.

1. $(G, *)$ is a groupoid. For any two elements $x, y \in G, x * y \in G$. If this condition is satisfied, we say that G is closed under $*$. If we can find two elements $x, y \in G$ such that $x * y \notin G$, then $(G, *)$ cannot be a groupoid and hence cannot be a group.
2. $*$ is associative. If $x, y, z \in G$, then $x * (y * z) = (x * y) * z$. If we can produce elements a, b, c for which $a * (b * c) \neq (a * b) * c$, then $*$ is not associative and $(G, *)$ is not a group.
3. There is an element e which has the property that $x * e = e * x = x \ \forall x \in G$. If we cannot find such an element e, then $(G, *)$ cannot be a group.
4. For any element $x \in G$, there corresponds an element $y \in G$ satisfying the condition $x * y = y * x = e$. If we can find one element in G for which there does not correspond such an element, then we conclude that $(G, *)$ is not a group.
5. If we further want to show that $(G, *)$ is commutative, we have to show that $x * y = y * x$ for all $x, y \in G$. If we can produce elements a and b in G for which $a * b \neq b * a$, we can conclude that $(G, *)$ is not commutative.

   A commutative group $(G, *)$ is also called an *abelian group*.

Result

Let $(G, *)$ be a group and $a, b \in G$. Then,

(i) $(a^{-1})^{-1} = a$. This means the inverse of a^{-1} is a.
(ii) $(a * b)^{-1} = b^{-1} * a^{-1}$.

Proof

(i) Let $x = a^{-1}$. We have to prove that $x^{-1} = a$, i.e., inverse of x is a. For this we have to prove that $x * a = a * x = e$. Now $x * a = a^{-1} * a = e$ and $a * x = a * a^{-1} = e$. This proves that $x^{-1} = a$.

(ii) Let $x = a * b$ and $y = b^{-1} * a^{-1}$. We have to prove that $x^{-1} = y$, i.e., inverse of x is y. Now $x * y = (a * b) * (b^{-1} * a^{-1}) = a * (b * b^{-1}) * a^{-1} = a * e * a^{-1} = a * a^{-1} = e$. Similarly, we can prove that $y * x = e$. Since $x * y = y * x = e$, it follows that $x^{-1} = y$.

If G contains only a small number of elements, we can express the binary operation $*$ on G in a very compact form using a table. This table is called the *multiplication table*. This should not be confused with ordinary multiplication of numbers.

The following examples show this way of defining binary operation.

1. Let $Z_4 = \{0, 1, 2, 3\}$ and define a binary operation $\oplus$ on Z_4 as follows.

$\oplus$	0	1	2	3
0	0	1	2	3
1	1	2	3	0
2	2	3	0	1
3	3	0	1	2

To add two elements of Z_4, we first add them like ordinary numbers. If this sum is less than 4, we write the sum as it is. Otherwise, we divide the sum by 4 and write the remainder. When the sum is divided by 4, the remainder is either 0 or 1 or 2 or 3. Hence Z_4 is closed under this operation. In other words, if $m, n \in Z_4$, then $m \oplus n \in Z_4$. (Note that all the elements in the table are elements belonging to Z_4). From the table it is clear that 0 is the identity element and every element of Z_4 has an inverse (the inverse of 0 is 0, inverse of 1 is 3, inverse of 2 is 2 and inverse of 3 is 1). From the table, it is clear that $\oplus$ operation is both associative and commutative. Hence, $(Z_4, \oplus)$ is a commutative group.

2. Let $Z_n = \{0, 1, 2, \ldots, n-1\}$ and let $\oplus$ be defined as follows. If $p, q \in Z_n$, then $p \oplus q = (p + q) \bmod n$. We add p and q as ordinary numbers and take the remainder when $p + q$ is divided by n. Then we can see that $(Z_n, \oplus)$ is a commutative group in which 0 is the identity element and inverse of any element m is $n - m$. We have $m \oplus (n - m) = 0$, which is the identity element.

3. Let $G = \{1, -1, i, -i\}$ and let $*$ be defined as follows.

$*$	1	-1	i	$-i$
1	1	-1	i	$-i$
-1	-1	1	$-i$	i
i	i	$-i$	-1	1
$-i$	$-i$	i	1	-1

Then $(G, *)$ is a commutative group whose identity element is 1.

If $(G, *)$ is a group, then in the multiplication table, in every row and in every column, all elements of G occur once and only once. This means that in every row we can encounter all elements of G. Also, no element occurs more than once in any row. It is likewise in every column. If an element appears more than once in a row or in a column or if an element does not appear in a row or in a column, then we can conclude that $(G, *)$ is not a group. In fact, we can prove the following.

Theorem

(i) In a group $(G, *)$, if $a * b = a * c$, then $b = c$.
(ii) In a group $(G, *)$, if $b * a = c * a$, then $b = c$.

Property (i) is called the *left-cancellation property*. The same element can be cancelled from the left side. Property (ii) is called the *right-cancellation property*. The same element can be cancelled from the right side.

Proof We will prove only (i). (ii) can be proved similarly.

We assume $a * b = a * c$ and we will prove that $b = c$. Since $(G, *)$ is a group, a has an inverse a^{-1}. Pre-multiplying both sides of $a * b = a * c$ by a^{-1}, we obtain $a^{-1} * (a * b) = a^{-1} * (a * c)$. Since $*$ is associative, this means $(a^{-1} * a) * b = (a^{-1} * a) * c$. By definition of a^{-1} we obtain $e * b = e * c$, which is the same as saying $b = c$ (by definition of identity element).

Subgroups

Definition Let $(G, *)$ be a group and let H be a subset of G. We say H is *a subgroup of* G if $(H, *)$ is a group, i.e., H is a group under the same binary operation as in G.

If $x, y \in H$, then $x, y \in G$ and hence $x * y \in G$. If H has to be a subgroup of G we require that $x * y \in H$. This means that H is closed under the binary operation $*$. Clearly, associativity holds in H since it holds in G. If H has to be a subgroup of G, then the identity element e of G must belong to H. If $x \in H$ then $x \in G$ and hence $x^{-1} \in G$. If H has to be a subgroup, we require that $x^{-1} \in H$. Thus, in order to prove that H is a subgroup of G, we have to prove the following.

(i) If $x, y \in H$, then $x * y \in H$.
(ii) $e \in H$.
(iii) If $x \in H$, then $x^{-1} \in H$.

The following theorem says that it is not necessary to verify all the above to prove that a non-empty subset H of G is a subgroup of G.

Theorem Let H be a non-empty subset of a group $(G, *)$. Then H is a subgroup of G if and only if $a * b^{-1} \in H \ \forall\, a, b \in H$.

Proof Suppose H is a subgroup of $(G, *)$. Then $(H, *)$ is a group. If $a, b \in H$, then $b^{-1} \in H$ and hence $a * b^{-1} \in H$.

Conversely, assume that $a * b^{-1} \in H \ \forall\, a, b \in H$. Since H is non-empty, H contains at least one element a. Since $a, a \in H$, by our assumption $a * a^{-1} \in H$, i.e., $e \in H$. Thus H contains the identity element of G. For any element $h \in H$, since $e, h \in H$, by our assumption $e * h^{-1} \in H$, i.e., $h^{-1} \in H \ \forall\, h \in H$. If $x, y \in H$, then $x, y^{-1} \in H$ and hence by our assumption $x * (y^{-1})^{-1} \in H$, i.e., $x * y \in H$. Associativity property is true in H because it is true in G. Thus $(H, *)$ satisfies all the properties of a group and hence $(H, *)$ is a subgroup of $(G, *)$.

Solved problems

1. Let $*$ be defined on Z as $a * b = a + b - ab$, where ab denotes the product of a and b. Determine whether $(Z, *)$ is a commutative group.

   Solution If a and b are integers, then $a + b - ab$ is also an integer. Hence $a * b \in Z$. This proves that Z is closed under $*$. If $a, b, c \in Z$, then $(a * b) * c = (a + b - ab) * c = a + b - ab + c - (a + b - ab)c = a + b - ab + c - ac - bc + abc$. Again, $a * (b * c) = a * (b + c - bc) = a + b + c - bc - a(b + c - bc) = a + b + c - bc - ab - ac + abc$. Hence $(a * b) * c = a * (b * c)$, proving that $*$ is associative. For every $a \in Z$, we have $a * 0 = a = 0 * a$, proving that 0 is the identity element of $(Z, *)$. If $1 * b = 0$ for some b, then $1 + b - 1 \times b = 0$. This means $1 = 0$, which is not true. Hence 1 does not have an inverse, showing that $(Z, *)$ is not a group.

2. Let $*$ be defined on $\mathbb{R} - \{-1\}$ as $a * b = a + b + ab$ (ab denotes the product of a and b). Determine whether $(\mathbb{R} - \{-1\}, *)$ is a commutative group.

   Solution We will first prove that if $a, b \in \mathbb{R} - \{-1\}$, then $a + b + ab \in \mathbb{R} - \{-1\}$. Clearly, $a + b + ab \in \mathbb{R}$. If $a + b + ab = -1$, then $b(1 + a) = -1 - a = -(1 + a)$. Since $a \neq -1$, $1 + a \neq 0$. Dividing both sides of $b(1 + a) = -(1 + a)$ by $(1 + a)$, we get $b = -1$, which is not true. Hence $a + b + ab \neq -1$ so that $a * b = a + b + ab \in \mathbb{R} - \{-1\}$. Hence $\mathbb{R} - \{-1\}$ is closed under $*$. It is easy to see that $*$ is associative. $a * 0 = 0 * a = a$ shows that 0 is the identity element. If $a * b = 0$, then $a + b + ab = 0$. Hence $b(1 + a) = -a$. Since $a \neq -1$, $1 + a \neq 0$ so that $b = \frac{-a}{(1+a)}$. Thus every element $a \in \mathbb{R} - \{-1\}$ has an inverse $\frac{-a}{(1+a)}$. Clearly, $a * b = b * a$. Hence $(\mathbb{R} - \{-1\}, *)$ is a commutative group.

3. Prove that a group G is commutative if and only if $\forall\, a, b \in G$, $(a*b)^2 = a^2 * b^2$.

   Solution Suppose G is commutative. Let $a, b \in G$. Since $a*b = b*a$, we have $(a*b)^2 = (a*b)*(a*b) = a*b*a*b = a*a*b*b = a^2*b^2$.

   Conversely, assume that $\forall a, b \in G$, $(a * b)^2 = a^2 * b^2$. Then $\forall\, a, b \in G$, $a * b * a * b = a * a * b * b$. By left-cancellation property, we obtain $b * a * b = a * b * b$. Again by right-cancellation property we obtain $b * a = a * b$, proving that G is commutative.

4. Let $(G, *)$ be a finite group with identity element e and let $a \in G$. Prove that there exists a positive integer n such that $a^n = e$.

   Solution Consider the elements $a, a^2 = a * a, a^3 = a * a * a \ldots a^m = a*a*a*\ldots$ (m times), ... All these elements belong to G (since G is closed under $*$). Since G contains only a finite number of elements, all these elements cannot be distinct. In other words, we can find positive integers i and j $(i \neq j)$ such that $a^i = a^j$. If $i > j$, then $a^{i-j} = a^i * a^{-j} = a^j * a^{-j} = e$. We take $n = i - j > 0$. If $j > i$, then $a^{j-i} = e$ and we can take $n = j - i > 0$.

5. If H and K are subgroups of a group $(G, *)$, prove that $H \cup K$ is a subgroup of $(G, *)$ if and only if either $H \subseteq K$ or $K \subseteq H$.

   Solution Suppose $H \subseteq K$ or $K \subseteq H$. If $H \subseteq K$, then $H \cup K = K$ and if $K \subseteq H$, then $H \cup K = H$. In both cases $H \cup K$ is a subgroup of $(G, *)$.

   Conversely, assuming that $H \cup K$ is a subgroup of $(G, *)$, we will prove that either $H \subseteq K$ or $K \subseteq H$. Let, if possible, $H \not\subseteq K$ and $K \not\subseteq H$. Then there exists an element $h \in H$ such that $h \notin K$. Also, there exists an element $k \in K$ such that $k \notin H$. Now consider the element hk (note that hk means $h * k$). Since $h \in H$ and $k \in K$, h and k both belong to $H \cup K$. Since $H \cup K$ is a subgroup, it follows that $hk \in H \cup K$. Hence $hk \in H$ or $hk \in K$. If $hk \in H$, then since $h^{-1} \in H$, we obtain $h^{-1}(hk) = (h^{-1}h)k = ek = k \in H$ contradicting the fact that $k \notin H$. If we assume $hk \in K$, we will get the contradiction $h \in K$. Hence our assumption that $H \not\subseteq K$ and $K \not\subseteq H$ is false. This proves that either $H \subseteq K$ or $K \subseteq H$.

6. Let $(S, *)$ be a semigroup and e be a left identity. Furthermore, assume that for every x in S, there exists x' in S such that $x' * x = e$. Show the following.

(a) For any a, b, c in S, if $a * b = a * c$, then $b = c$.
(b) $(S, *)$ is a group.

Solution To prove (a), assume $a * b = a * c$. By assumption, there exists an a' in S such that $a' * a = e$. Pre-multiplying both sides by a', we obtain $a' * (a * b) = a' * (a * c)$, which by associativity means $(a' * a) * b = (a' * a) * c$. Since $a' * a = e$, it follows that $e * b = e * c$, which means $b = c$ (since e is a left identity).

To prove (b), we will first prove that e is also a right identity. For this we have to prove that for every $x \in S, x * e = x$. By assumption there exists $x' \in S$ such that $x' * x = e$. Now $x' * (x * x') * x = (x' * x) * (x' * x) = e * e = e = x' * x$. By (a) we obtain $(x * x') * x = x$. By associativity this means $x * (x' * x) = x$. Since $x' * x = e$, it follows that $x * e = x$. It remains to prove that x' is also a right inverse of x. By associativity and using the fact that $x' * x = e$, we obtain $(x * x') * (x * x') = x * (x' * x) * x' = x * e * x' = x * x'$ (since e is identity element). This means $x * (x' * (x * x')) = x * x'$. By left-cancellation property (we have already proved (a)), we obtain $x' * (x * x') = x' = x' * e$ (since e is right identity). Again, by left-cancellation property we obtain $x * x' = e$. This completes the proof.

7. Let $(G, *)$ be a group. Show that $(G, *)$ is an abelian group if $a^3 * b^3 = (a * b)^3$, $a^4 * b^4 = (a * b)^4$ and $a^5 * b^5 = (a * b)^5$ for all a and b in G.

Solution For the sake of simplicity, we will denote $a * b$ by ab. Let $a, b \in G$. We will prove that $ab = ba$. This will prove that G is commutative. We are given the following.

$$aaabbb = ababab \tag{1}$$

$$aaaabbbb = abababab \tag{2}$$

$$aaaaabbbbb = ababababab \tag{3}$$

Substituting (1) in (2), we obtain $aabababb = abababab$.
By right-cancellation property, we have

$$aababab = abababa. \tag{4}$$

Again, substituting (2) in (3) we obtain

$$aababababb = ababababab. \tag{5}$$

Substituting (4) in (5), we obtain $abababaabb = ababababab$.
Repeatedly applying left-cancellation and right-cancellation properties, we finally obtain $ab = ba$.

8. Let $(H, *)$ and $(K, *)$ be subgroups of a group $(G, *)$. Let $HK = \{h * k | h \in H \text{ and } k \in K\}$. Show that $(HK, *)$ is a subgroup of $(G, *)$ if and only if $HK = KH$.

Solution Again we will omit $*$ and denote $a * b$ by ab.
We will first assume that HK is a subgroup of G and we will prove that $HK = KH$. Let $hk \in HK (h \in H, k \in K)$. Since HK is a subgroup of G, $(hk)^{-1} = k^{-1}h^{-1} \in HK$. Let $k^{-1}h^{-1} = h_1k_1 (h_1 \in H, k_1 \in K)$. Then $hk = (k^{-1}h^{-1})^{-1} = (h_1k_1)^{-1} = k_1^{-1}h_1^{-1} \in KH$ (note that $k_1^{-1} \in K, h_1^{-1} \in H$ since H and K are subgroups). We have thus proved that $HK \subseteq KH$. Similarly, we can prove that $KH \subseteq HK$. Hence $HK = KH$.

Conversely, assuming that $HK = KH$, we will prove that HK is a subgroup of G. Since $e \in H$ and $e \in K$, $e = ee \in HK$ proving that HK is non-empty. Let $hk \in HK$ and $h_1k_1 \in HK (h, h_1 \in H, k, k_1 \in K)$. We have $(hk)(h_1k_1)^{-1} = (hk)(k_1^{-1}h_1^{-1}) = h(kk_1^{-1})h_1^{-1} = hk_2h_1^{-1}$, where $k_2 = kk_1^{-1} \in K$. Now $k_2h_1^{-1} \in KH = HK$ (by assumption). Let $k_2h_1^{-1} = h'k' (h' \in H, k' \in K)$. Then $(hk)(h_1k_1)^{-1} = hk_2h_1^{-1} = hh'k' \in HK$ (note that $h, h' \in H$ imply $hh' \in H$ since H is a subgroup). We have thus proved that HK is a subgroup of G.

Exercises

1. Determine whether each of the sets given below together with the given operation is a semigroup, monoid or neither. If it is a semigroup or monoid, determine whether it is commutative.

   (i) $Z^+, a * b = \max\{a, b\}$.
   (ii) $Z^+, a * b = \gcd\{a, b\}$.
   (iii) $Z^+, a * b = a$.
   (iv) $(P(S), \cap)$ where S is any set.
   (v) $S = \{1, 2, 3, 6, 12\}, a * b = \gcd\{a, b\}$.

2. Prove that the intersection of two subgroups is again a subgroup.
3. Determine whether each of the sets given below together with the given operation is a commutative group.

   (i) $(\mathbb{R}, *)$ where $a * b = a + b + 2$.
   (ii) $(P(S), *)$ where $A * B = A \oplus B$.

4. Let $S = \{x \in \mathbb{R} | x \neq 0 \text{ and } x \neq 1\}$. Consider the following functions defined from S to itself.

   $f_1(x) = x$, $f_2(x) = (1 - x)$, $f_3(x) = \frac{1}{x}$, $f_4(x) = \frac{1}{(1-x)}$, $f_5(x) = 1 - \frac{1}{x}$ and $f_6(x) = \frac{x}{(x-1)}$.

Prove that $G = \{f_1, f_2, f_3, f_4, f_5, f_6\}$ is a group under the operation of composition.

5. Let $(G, *)$ be a group. If $x * x = x$, prove that $x = e$.
6. Prove that if $a^2 = e \;\forall\; a \in G$, then G is commutative.
7. Let $(G, *)$ be a group. If $a * b = b * a$, prove that $(a * b)^n = a^n * b^n$ for all positive integers n.

   Note For a positive integer n, $a^n = a * a * a * \ldots n$ times and $a^{-n} = a^{-1} * a^{-1} * a^{-1} * \ldots n$ times.
8. Let $(S, *)$ be a semigroup. Furthermore, let there be an element s in S such that for every x in S there exist u and v in S satisfying the relation

$$s * u = v * s = x.$$

   Show that $(S, *)$ contains an identity element.
9. Let H_1 and H_2 be subgroups of a group G neither of which contains the other. Show that there exists an element of G belonging to neither H_1 nor H_2.
10. Let $(H, *)$ be a subgroup of a group $(G, *)$. Let $N = \{x | x \in G \text{ and } xHx^{-1} = H\}$. Show that $(N, *)$ is a subgroup of $(G, *)$.

Let $(G, *)$ be a group. If $a, b \in G$, we will denote $a * b$ by ab. The notation ab does not have anything to do with the product of a and b. It is assumed there is a binary operation between a and b which is omitted for the sake of convenience.

If a group $(G, *)$ contains a finite number of elements, it is called a finite group and the number of elements in G is called the order of G. It is denoted by $|G|$.

Let S_n denote the set of all permutations on n symbols $a_1, a_2, \ldots, a_n$. Every element of S_n is a function from the set $\{a_1, a_2, \ldots, a_n\}$ to itself which is both one–one and onto. Then $(S_n, \circ)$ is a finite group where '$\circ$' denotes the composition of permutations. Its order is $n!$.

If $n = 3$ then the elements of S_3 are the following.

$$f_1 = \begin{matrix} a_1 & a_2 & a_3 \\ a_1 & a_2 & a_3 \end{matrix} \qquad f_2 = \begin{matrix} a_1 & a_2 & a_3 \\ a_1 & a_3 & a_2 \end{matrix} \qquad f_3 = \begin{matrix} a_1 & a_2 & a_3 \\ a_3 & a_2 & a_1 \end{matrix}$$

$$f_4 = \begin{matrix} a_1 & a_2 & a_3 \\ a_2 & a_1 & a_3 \end{matrix} \qquad f_5 = \begin{matrix} a_1 & a_2 & a_3 \\ a_2 & a_3 & a_1 \end{matrix} \qquad f_6 = \begin{matrix} a_1 & a_2 & a_3 \\ a_3 & a_1 & a_2 \end{matrix}$$

The order of S_3 is $3! = 6$. The identity element of S_3 is f_1.

Problem Find the inverse of various elements of S_3.

Examples

1. $(A_n, \circ)$ is a subgroup of $(S_n, \circ)$ where A_n denotes the set of all even permutations. This subgroup is called the alternating group on n symbols.
2. If O_n denotes the set of all odd permutations, then $(O_n, \circ)$ is not a subgroup of $(S_n, \circ)$. If $p, q \in O_n$ then $p \circ q \notin O_n$. In fact, $p \circ q$ is an even permutation.

Result Let $(S_1, *_1)$ and $(S_2, *_2)$ be two semigroups. Define a binary operation $*$ on $S_1 \times S_2$ as follows.

If $a, a_1 \in S_1$ and $b, b_1 \in S_2$, then $(a, b) * (a_1, b_1) = (a *_1 a_1, b *_2 b_1)$. Then $(S_1 \times S_2, *)$ is a semigroup.

Proof Since $a, a_1 \in S_1, a *_1 a_1 \in S_1$ (since $*_1$ is a binary operation on S_1). Similarly $b *_2 b_1 \in S_2$. Hence $(a, b) * (a_1, b_1) = (a *_1 a_1, b *_2 b_1) \in S_1 \times S_2$. This proves that $(S_1 \times S_2, *)$ is a groupoid. If $(a, b), (a_1, b_1), (a_2, b_2) \in S_1 \times S_2$, then

$$
\begin{aligned}
[(a, b) * (a_1, b_1)] * (a_2, b_2) &= (a *_1 a_1, b *_2 b_1) * (a_2, b_2) \\
&= ((a *_1 a_1) *_1 a_2, (b *_2 b_1) *_2 b_2)) \\
&= (a *_1 (a_1 *_1 a_2), b *_2 (b_1 *_2 b_2)) \quad \text{(by associativity of } *_1 \text{ and } *_2\text{).} \\
&= (a, b) * (a_1 *_1 a_2, b_1 *_2 b_2) \\
&= (a, b) * [(a_1, b_1) * (a_2, b_2)].
\end{aligned}
$$

This proves that $*$ is associative and hence $(S_1 \times S_2, *)$ is a semigroup.

Result If $(M_1, *_1)$ and $(M_2, *_2)$ are monoids, then $(M_1 \times M_2, *)$ is also a monoid where $*$ is defined as above.

Proof Let e_1 and e_2 denote the identity elements of M_1 and M_2 respectively. (We know e_1 and e_2 exist since $(M_1, *_1)$ and $(M_2, *_2)$ are monoids.) Then for any element $(m_1, m_2) \in M_1 \times M_2$, we have $(m_1, m_2) * (e_1, e_2) = (m_1 *_1 e_1, m_2 *_2 e_2) = (m_1, m_2)$. Similarly, $(e_1, e_2) * (m_1, m_2) = (m_1, m_2)$. This proves that (e_1, e_2) is the identity element of $M_1 \times M_2$.

Result If $(G_1, *_1)$ and $(G_2, *_2)$ are groups, then $(G_1 \times G_2, *)$ is also a group where $*$ is as defined above.

Proof We already know that $(G_1 \times G_2, *)$ is a monoid. Let $(a, b) \in G_1 \times G_2$ where $a \in G_1$ and $b \in G_2$. Since $(G_1, *_1)$ is a group, a has an inverse a^{-1} belonging to G_1. Similarly, b has an inverse b^{-1} belonging to G_2. Hence $(a^{-1}, b^{-1}) \in G_1 \times G_2$ and $(a, b) * (a^{-1}, b^{-1}) = (a *_1 a^{-1}, b *_2 b^{-1}) = (e_1, e_2)$, where e_1 is the identity element of $(G_1, *_1)$ and e_2 is the identity element of $(G_2, *_2)$. Similarly, $(a^{-1}, b^{-1}) * (a, b) = (e_1, e_2)$, which is the identity element of $G_1 \times G_2$.

The group $(G_1 \times G_2, *)$ is called the product of the groups $(G_1, *_1)$ and $(G_2, *_2)$.

Let $(S, *)$ be a semigroup and let R be an equivalence relation on the set S. R is said to be a congruence relation if the following condition is satisfied.

Whenever $a\ R\ b$ and $c\ R\ d (a, b, c, d \in S)$, then $a * c\ R\ b * d$. In other words, whenever $(a, b) \in R$ and $(c, d) \in R (a, b, c, d \in S)$, then $(a * c, b * d) \in R$.

Example Consider the group $(Z, +)$ and the relation R defined as $a\ R\ b$ if and only if $4|(a - b)$. Then R is a congruence relation.

Proof Clearly, R is an equivalence relation. Suppose $a\ R\ b$ and $c\ R\ d$. We will prove that $(a + c)\ R\ (b + d)$. Since $a\ R\ b$, we have $4|(a - b)$. Hence $a - b = 4k$. Similarly, $c\ R\ d$ implies $c - d = 4k_1 (k, k_1 \in Z)$. Adding, we obtain $(a - b) + (c - d) = (a + c) - (b + d) = 4k + 4k_1 = 4(k + k_1)$, where $k + k_1 \in Z$. This proves that $4|((a + c) - (b + d))$ and hence $(a + c)\ R\ (b + d)$.

Generalising the above, we obtain the following.

Consider the group $(Z, +)$ and the relation R defined as $a\ R\ b$ if and only if $n|(a - b)$, where n is any positive integer. Then R is a congruence relation on $(Z, +)$.

Result Let $(S, *)$ be a semigroup and let R be a congruence relation on $(S, *)$. Consider the quotient set $S|R$. Define a binary operation on $S|R$ as $[a] \circledast [b] = [a * b]$. Then $(S|R, \circledast)$ is a semigroup.

Proof First we have to prove that this operation $\circledast$ is well defined. This is very essential. Now a is only one element of $[a]$ and b is only one element of $[b]$. If we take another element c from $[a]$ and another element d from $[b]$, we can also define $[a] \circledast [b] = [c * d]$. In that case we have to prove that $[a * b] = [c * d]$. This will prove that the definition of $\circledast$ is independent of the elements we choose from $[a]$ and $[b]$.

Since $c \in [a]$, we have $a\ R\ c$. Since $d \in [b]$, we have $b\ R\ d$. By the definition of congruence relation it follows that $a * b\ R\ c * d$, proving that $[a * b] = [c * d]$.

Since $a, b \in S$ implies $a * b \in S$, we have $[a] \circledast [b] = [a * b] \in S|R$. Hence $S|R$ is closed under $\circledast$.

$\circledast$ is associative since

$$\begin{aligned}([a] \circledast [b]) \circledast [c] &= [a * b] \circledast [c] = [(a * b) * c] = [a * (b * c)] \\ &= [a] \circledast [b * c] = [a] \circledast ([b] \circledast [c]).\end{aligned}$$

The semigroup $(S|R, \circledast)$ is called the quotient semigroup.

Consider the group $(Z, +)$. We know that every positive integer can be expressed using only 1 and +. For example, $5 = 1 + 1 + 1 + 1 + 1$. In general, if n is any positive integer, then $n = 1 + 1 + 1 + \ldots + 1$ (n times). Similarly, any negative integer $-m$ (m is a positive integer) can be expressed as $-1 + -1 + -1 \ldots + -1$ (m times). 1 and -1 are called generators of the group $(Z, +)$ since any element of $(Z, +)$ can be generated using 1 and -1.

A group $(G, *)$ is said to be a cyclic group if it can be generated by one element. In other words, in a cyclic group $(G, *)$ there exists an element a such that every element $x \in G$ can be expressed as $x = a^n$ for some positive integer n. Note that there can also be another element b such that every element $y \in G$ can be expressed as $y = b^m$ for some positive integer m.

Example $(Z, +)$ is not a cyclic group since it is generated by two elements 1 and -1.

Result A cyclic group $(G, *)$ is commutative.

Proof Let $(G, *)$ be a cyclic group and let a be a generator of $(G, *)$. If $x, y \in G$, then $x = a^n$ and $y = a^m$ for some positive integers m and n. Now,

$$\begin{aligned} x * y = a^n * a^m &= a * a * \ldots * a (n \text{ times}) * a * a * \ldots * a (m \text{ times}) \\ &= a * a * \ldots * a (m + n \text{ times}) \\ &= a^{m+n}. \end{aligned}$$

Similarly, $y * x = a^{m+n}$. Hence $(G, *)$ is commutative .

Consider two semigroups $(S, *)$ and $(S_1, *_1)$. A function $f : S \to S_1$ is said to be a homomorphism if $f(a*b) = f(a) *_1 f(b)$ for all $a, b \in S$. Further, if f is one–one, it is called a monomorphism. Thus a monomorphism is a homomorphism which is one–one. A homomorphism which is onto is called an epimorphism. If a homomorphism is both one–one and onto, it is called an isomorphism.

Consider two groups $(G, *)$ and $(G_1, *_1)$. Since $(G, *)$ and $(G_1, *_1)$ are also semigroups, we can define homomorphism and isomorphism between the groups $(G, *)$ and $(G_1, *_1)$. If there is an isomorphism between the groups $(G, *)$ and $(G_1, *_1)$, then the groups $(G, *)$ and $(G_1, *_1)$ are said to be isomorphic. If two groups $(G, *)$ and $(G_1, *_1)$ are isomorphic, then they share all the interesting properties. If we want to show that two groups $(G, *)$ and $(G_1, *_1)$ are not isomorphic, we will have to identify at least one property which one of the groups possesses but the other does not.

If two finite groups are isomorphic, then the number of elements in the two groups must be the same. Hence two finite groups whose orders are different cannot be isomorphic. If two groups contain the same number of elements, then the two groups may or may not be isomorphic. For example, the groups $(Z_6, \oplus)$ and $(S_3, \circ)$ cannot be isomorphic even though both of them have the same number of elements. This is because $(Z_6, \oplus)$ is commutative but $(S_3, \circ)$ is not commutative.

Examples

1. Let G denote the group of all real numbers with binary operation of addition and G_1 the group of all positive real numbers with binary operation of multiplication. Define $f : G \to G_1$ as $f(x) = e^x$. Then f is a homomorphism because $f(x+y) = e^{x+y} = e^x e^y = f(x)f(y)$. f is one–one since $f(x) = f(y)$ implies $e^x = e^y$, which means $x = y$. To prove f is onto, consider $x \in \mathbb{R}^+$. We have to find $y \in \mathbb{R}$ such that $f(y) = x$. Take $y = \ln x$. Then $y \in \mathbb{R}$ and $f(y) = e^y = e^{\ln x} = x$. Since f is a homomorphism, one–one and onto, it follows that f is an isomorphism. Hence the groups G and G_1 are isomorphic.
2. Let $G = (Z, +)$ and $G_1 = (Z_n, \oplus)$. Define $f : Z \to Z_n$ as $f(m) = r$, where r is the remainder obtained when m is divided by n. Then f is an onto homomorphism.

Proof Consider $k \in Z_n$. Then $0 \le k \le n-1$. Now $f(k) = k$, proving that f is onto.

To prove f is a homomorphism, we will prove $f(k+l) = f(k) \oplus f(l)$ for all $k, l \in Z$.

Suppose $k = q_1 n + r_1, 0 \le r_1 \le n-1$ and $l = q_2 n + r_2,\ 0 \le r_2 \le n-1$. By definition of f, we have $f(k) = r_1,\ f(l) = r_2$. Also, $k+l = (q_1+q_2)n + (r_1+r_2)$. Let $r_1 + r_2 = q_3 n + r_3, 0 \le r_3 \le n-1$. Then $k + l = (q_1+q_2)n + q_3 n + r_3 = (q_1+q_2+q_3)n + r_3$ and hence $f(k+l) = r_3$. Also, $f(k) \oplus f(l) = r_1 \oplus r_2 = r_3$, proving that f is a homomorphism.

Result Let $(M, *)$ and $(M_1, *_1)$ be two monoids with identity elements e and e_1 respectively and let $f : M \to M_1$ be an onto homomorphism. Then $f(e) = e_1$.

Proof We will prove that $f(e)$ is the identity element of $(M_1, *_1)$. Since identity element in a monoid is unique, it will follow that $f(e) = e_1$. Consider any element $m_1 \in M_1$. Since f is onto, there exists $m \in M$ such that $f(m) = m_1$. Now $m_1 *_1 f(e) = f(m) *_1 f(e) = f(m * e) = f(m) = m_1$. Similarly, $f(e) *_1 m_1 = m_1$, proving that $f(e)$ is the identity element of $(M_1, *_1)$.

Result Let $(G, *)$ and $(G_1, *_1)$ be two groups with identity elements e and e_1 respectively and let $f : G \to G_1$ be a homomorphism. Then $f(e) = e_1$ and for all $a \in G$, $f(a^{-1}) = f(a)^{-1}$.

Note

1. In this case, it is not necessary that the homomorphism f is onto.
2. a^{-1} is the inverse of a in the group $(G, *)$ and $f(a)^{-1}$ is the inverse of $f(a)$ in the group $(G_1, *_1)$.

Proof We will first prove that $f(e) = e_1$. Let $x = f(e)$. Now $x *_1 x = f(e) *_1 f(e) = f(e * e) = f(e) = x$ from which it follows that $x = e_1$.

Next we have to prove that the inverse of $f(a)$ is $f(a^{-1})$. For this it suffices to prove that $f(a^{-1}) *_1 f(a) = e_1 = f(a) *_1 f(a^{-1})$. Since f is a homomorphism, we have $f(a^{-1}) *_1 f(a) = f(a^{-1} * a) = f(e) = e_1$. Similarly, $f(a) *_1 f(a^{-1}) = e_1$.

Definition Consider two monoids $(M, *)$ and $(M_1, *_1)$. Let $f : M \to M_1$ be a homomorphism. We define the kernel of f denoted by ker (f) as

$$\ker(f) = \{x \in M \mid f(x) = e_1\}.$$

It is the set of all elements of M which are mapped onto the identity element of M_1.

Result Let $(G, *)$ and $(G_1, *_1)$ be two groups and $f : G \to G_1$ be a homomorphism. Then,

(i) ker (f) is a subgroup of G,
(ii) f is a monomorphism if and only if ker $(f) = \{e\}$.

Proof

(i) We first note that ker (f) is non-empty since $e \in$ ker (f). (Note that $f(e) = e_1$.) To prove ker (f) is a subgroup, assume that $x, y \in$ ker (f). Then $f(x) = f(y) = e_1$. We will prove that $x * y^{-1} \in$ ker (f). Since f is a homomorphism, it follows that $f(x * y^{-1}) = f(x) *_1 f(y^{-1}) = f(x) *_1 f(y)^{-1} = e_1 *_1 e_1^{-1} = e_1$, proving that $x * y^{-1}$ ker (f).
(ii) Assume f is a monomorphism. Suppose $x \in$ ker (f). Then by definition of ker (f), $f(x) = e_1 = f(e)$. Since f is one–one, it follows that $x = e$, proving that ker $(f) = \{e\}$.

Conversely, assume that ker $(f) = \{e\}$. We will prove that f is one–one. Suppose $f(x) = f(y)$. Since f is a homomorphism, we have $f(x * y^{-1}) = f(x) *_1 f(y^{-1}) = f(x) *_1 f(y)^{-1} = f(x) *_1 f(x)^{-1} = e_1$. By definition of ker (f), $x * y^{-1} \in$ ker $(f) = \{e\}$. Hence $x * y^{-1} = e$. Post-multiplying both sides by y, we obtain $(x * y^{-1}) * y = e * y$. By associativity, this reduces to $x * (y^{-1} * y) = e * y$, i.e., $x * e = e * y$, which means $x = y$. Hence f is one–one.

Definition Let $(G, *)$ be a group and H be a subgroup of G. If $a \in G$, we define the left coset of H in G denoted by aH as $aH = \{ah | h \in H\}$.

Note

1. ah actually means $a * h$.
2. We have $a = ae \in aH$ since $e \in H$. Hence aH is non-empty for any $a \in G$.

The right coset of H in G denoted by $Ha (a \in G)$ is defined as $Ha = \{ha | h \in H\}$. Again, $a \in Ha$ and hence Ha is non-empty for any $a \in G$. We note that it is not necessary that for an element $a \in G, aH = Ha$. However, if $aH = Ha$ for all $a \in G$, then H is called a normal subgroup of G.

Result Every subgroup of a commutative group is normal.

Proof Let $(G, *)$ be a commutative group and let H be a subgroup of G. We will prove that $aH = Ha$ for all $a \in G$. If $ah \in aH (h \in H)$, then by commutativity, $ah = ha \in Ha$. Hence $aH \subseteq Ha$. Similarly, we can prove that $Ha \subseteq aH$. Hence $aH = Ha$, proving that H is normal.

Lagrange's theorem

Let $(G, *)$ be a finite group and H be a subgroup of G. Then $\mathrm{o}(H) | \mathrm{o}(G)$, where $\mathrm{o}(H)$ denotes the order of H and $\mathrm{o}(G)$ denotes the order of G.

Proof Consider the collection of all left cosets of H in G i.e, $\{aH | a \in G\}$. We will first prove that this collection partitions G. For this, we have to prove the following.

(i) aH is a non-empty subset of $G \;\forall\; a \in G$.
(ii) $\cup_{a \in G} aH = G$.
(iii) For any two left cosets aH and bH, either $aH = bH$ or $aH \cap bH = \phi$.

Proof

(i) Consider any element $a \in G$. Since H is a subgroup of $G, e \in H$ and hence $ae \in aH$, i.e., $a \in aH$. This proves that aH is non-empty. If $ah \in aH (h \in H)$, then clearly $ah \in G$, proving that $aH \subseteq G$.

(ii) Since $aH \subseteq G\ \forall\ a \in G$, it follows that $\cup_{a\in G} aH \subseteq G$. To prove $G \subseteq \cup_{a\in G} aH$, take any element $g \in G$. Now $g \in gH \subseteq \cup_{a\in G} aH$, proving that $G \subseteq_{a\in G} aH$.

(iii) Assuming that $aH \cap bH \neq \phi$, we will prove that $aH = bH$. Since $aH \cap bH \neq \phi$, there is at least one element $x \in aH \cap bH$. Then $x = ah_1 = bh_2 (h_1, h_2 \in H)$. Post-multiplying by h_1^{-1}, we obtain $(ah_1)h_1^{-1} = (bh_2)h_1^{-1}$, which by associativity means $a(h_1 h_1^{-1}) = b(h_2 h_1^{-1})$. This means $ae = bh_3$, where $h_3 = h_2 h_1^{-1} \in H$. Since e is the identity element, it follows that $a = bh_3$. Now assume that $p \in aH$ so that $p = ah (h \in H)$. Substituting $a = bh_3$, we obtain $p = (bh_3)h = b(h_3 h) = bh_4$ where $h_4 = h_3 h \in H$. Hence $p \in bH$, proving that $aH \subseteq bH$. Similarly, we can prove that $bH \subseteq aH$.

Hence $o(G) = \sum_{a\in G} o(aH)$.

We will now prove that for every $a \in G$, $o(aH) = o(H)$.

Define a function $f : H \to aH$ as $f(h) = ah (h \in H)$.

If $f(h_1) = f(h_2) (h_1, h_2 \in H)$, then $ah_1 = ah_2$ and hence $h_1 = h_2$ (by left-cancellation property). This proves that f is one–one.

If $ah \in aH (h \in H)$, then $ah = f(h)$ proving that f is onto.

Since f is both one–one and onto , it follows that $o(H) = o(aH)$.

Let k denote the number of left cosets of H in G. k is the number of elements in $\{aH | a \in G\}$. Now,

$$o(G) = \sum_{a\in G} o(aH) = \sum o(H) \quad \text{(where the summation is taken } k \text{ times)}.$$

$$= k \times o(H).$$

This proves that $o(H) | o(G)$.

Applications of Lagrange's theorem

(i) If $o(G) = p$ where p is a prime number, then the only subgroups of G are $\{e\}$ and G itself. In other words, G does not have any non-trivial subgroups.

(ii) If $o(G) = 10$, then G cannot have subgroups of orders 3, 4, 6, 7, 8, 9 since none of these numbers divides 10.

Result Let $(G, *)$ be a group and H be a subgroup of G. Then H is a normal subgroup of G if and only if $aha^{-1} \in H\ \forall\ a \in G$ and $h \in H$.

Proof First assume that H is a normal subgroup of G. Let $a \in G$ and $h \in H$. By definition of normal subgroup, we have $aH = Ha$. Now $ah \in aH = Ha$ so that $ah = h_1a(h_1 \in H)$. Post-multiplying both sides by a^{-1} we obtain,

$$aha^{-1} = h_1aa^{-1} = h_1e = h_1 \in H.$$

Conversely, assume that $aha^{-1} \in H \; \forall \; a \in G$ and $h \in H$. We will prove that $aH = Ha \; \forall \; a \in G$. Let $ah \in aH$, where $h \in H$. By assumption, $aha^{-1} = h_1 \in H$. Post-multiplying both sides by a, we obtain $(aha^{-1})a = h_1a$. By associativity, we have $ah(a^{-1}a) = h_1a$, i.e., $ah(e) = h_1a$, which means $ah = h_1a \in Ha$, proving that $aH \subseteq Ha$. Similarly, we can prove that $Ha \subseteq aH$. Hence $aH = Ha$.

Fundamental homomorphism theorem for groups

Statement Let $(G, *)$ and $(G_1, *_1)$ be two groups and $f : (G, *) \to (G_1, *_1)$ be an onto homomorphism. Then ker (f) is a normal subgroup of $(G, *)$ and the quotient group $G|\ker(f)$ is isomorphic to $(G_1, *_1)$.

Proof We have already proved that ker (f) is a subgroup of $(G, *)$. We will now prove that ker (f) is a normal subgroup. Let $a \in G$ and $x \in \ker(f)$. Then $f(x) = e_1$. Since f is a homomorphism, we have $f(axa^{-1}) = f(a)f(x)f(a^{-1}) = f(a)f(x)f(a)^{-1} = f(a)e_1f(a)^{-1} = f(a)f(a)^{-1} = e_1$, proving that $axa^{-1} \in$ ker (f). Hence ker (f) is a normal subgroup of $(G, *)$.

For the sake of simplicity, we take $K = \ker(f)$ and we will prove that $G|K$ is isomorphic to $(G_1, *_1)$. Define a function $g : G|K \to G_1$ as $g(aK) = f(a)$, where $a \in G$.

We have to first prove that the function g is well defined. For this we have to prove that if $aK = bK$, then $f(a) = f(b)$. Now $b \in bK = aK$. Hence $b = ak(k \in K)$. Now $f(ab^{-1}) = f(a)f(b^{-1}) = f(a)f(k^{-1}a^{-1}) = f(a)f(k^{-1})f(a^{-1}) = f(a)e_1f(a^{-1})(k \in K$ implies $k^{-1} \in K$ and hence $f(k^{-1}) = e_1) = f(a)f(a^{-1}) = e_1$. We have thus proved that $e_1 = f(a)f(b^{-1}) = f(a)f(b)^{-1}$. Post-multiplying both sides by $f(b)$, we obtain $e_1f(b) = f(a)f(b)^{-1}f(b)$, i.e., $f(b) = f(a)e_1 = f(a)$.

Since f is a homomorphism, we have $g((aK)(bK)) = g((ab)K) = f(ab) = f(a)f(b) = g(aK)g(bK)$ proving that g is a homomorphism.

If $g(aK) = g(bK)$, then $f(a) = f(b)$. Now $f(ab^{-1}) = f(a)f(b^{-1}) = f(a)f(b)^{-1} = f(a)f(a)^{-1} = e_1$ so that $ab^{-1} \in K$. Let $ab^{-1} = k \in K$. Post-multiplying both sides by b, we obtain $(ab^{-1})b = kb$, i.e., $a(b^{-1}b) = kb$ which means $ae = a = kb \in Kb = bK$ since K is a normal subgroup. Hence $a \in aK \cap bK$ so that $aK = bK$. This proves that g is one–one.

Let $y \in G_1$. Since f is onto, there exists $x \in G$ such that $f(x) = y$. Now $xK \in G|K$ and $g(xK) = f(x) = y$. This proves that g is onto.

Solved problems

1. Let $(G, *)$ be a group and let $H = \{x \in G | xy = yx \text{ for all } y \in G\}$. Prove that H is a normal subgroup of G.

   Solution We will first prove that H is a subgroup of G. We first note that $e \in H$ because $ey = ye$ for all $y \in G$. Hence $H \neq \phi$. To prove that H is a subgroup of G, it is enough if we prove that $hk^{-1} \in H$ whenever $h, k \in H$. Let $y \in G$. Then $y^{-1} \in G$ so that $ky^{-1} = y^{-1}k$ (since $k \in H$). Since $h \in H$, we have $hy = yh$. Taking inverses on both sides of $ky^{-1} = y^{-1}k$, we obtain $yk^{-1} = k^{-1}y$ so that $(hk^{-1})y = h(k^{-1}y) = h(yk^{-1}) = (hy)k^{-1} = (yh)k^{-1} = y(hk^{-1})$, proving that $hk^{-1} \in H$. If $g \in G$ and $h \in H$, we will prove that $ghg^{-1} \in H$. Let $y \in G$. Since $h \in H$, by definition of H, $gh = hg$ so that $(ghg^{-1})y = hgg^{-1}y = hy = yh$ (since $h \in H$). Again, $y(ghg^{-1}) = y(hgg^{-1}) = yh$. Hence $ghg^{-1} \in H$, proving that H is a normal subgroup of G.

2. Let $(H, *)$ and $(K, *)$ be subgroups of a group $(G, *)$. Show that the function f from $H \times K$ to G defined as $f((h, k)) = h * k$ for all h in H and k in K is an isomorphism if and only if the following are satisfied.

   (a) H and K are normal subgroups.
   (b) $\{h * k | h \in H, k \in K\} = G$.
   (c) $H \cap K = \{e\}$ where e is the identity element.

   Solution For the sake of simplicity, we will denote $h * k$ by hk. We first assume that f is an isomorphism and we will prove (a), (b) and (c). To prove (a), we will only prove that H is a normal subgroup of G. The fact that K is a normal subgroup of G can be proved exactly similarly.

   To prove that H is a normal subgroup of G, assume that $g \in G$ and $h \in H$. We have to prove that $ghg^{-1} \in H$. Since $f : H \times K \to G$ is onto, there exist $h_1 \in H$ and $k_1 \in K$ such that $f((h_1, k_1)) = g$. By definition of f, we have $h_1k_1 = g$. Since f is a homomorphism and $e \in K$, $f((h, k_1)(h, e)) = f((h, k_1))f((h, e))$, i.e., $f((hh, k_1)) = hk_1h$. Again by the definition of f, we have $hhk_1 = hk_1h$. Cancelling h on left side, we obtain $hk_1 = k_1h$. Now $ghg^{-1} = h_1k_1hk_1^{-1}h_1^{-1} = h_1hk_1k_1^{-1}h_1^{-1} = h_1hh_1^{-1} \in H$. This proves that H is a normal subgroup of G.

   To prove (b), we first note that $\{hk | h \in H, k \in K\} \subseteq G$. (After all, every element belongs to G!) To prove that $G \subseteq \{hk | h \in H, k \in K\}$, assume

$g \in G$. Since $f : H \times K \to G$ is onto, there exist $h \in H$ and $k \in K$ such that $f((h, k)) = g$, i.e., $hk = g$, proving that $g \in \{hk | h \in H, k \in K\}$.

To prove (c), assume that $x \in H \cap K$. Then $x \in H$ and $x^{-1} \in K$ (since K is a subgroup) and $f((x, x^{-1})) = xx^{-1} = e$, proving that $(x, x^{-1}) \in \ker(f) = \{(e, e)\}$ since f is a monomorphism. Hence $x = e$ and this proves that $H \cap K = \{e\}$.

We will now assume (a), (b) and (c) and prove that f is an isomorphism. To prove f is onto, assume that $g \in G$. By (b), $g \in \{hk | h \in H, k \in K\}$ so that $g = hk$ for some $h \in H$ and $k \in K$. By definition of f, we have $g = f((h, k))$. This proves that f is onto.

To prove f is one–one, assume that $f((h_1, k_1)) = f((h_2, k_2))$. By definition of f, we have $h_1k_1 = h_2k_2$. Pre-multiplying both sides by h_2^{-1}, we obtain $h_2^{-1}h_1k_1 = k_2$. Post-multiplying both sides by k_1^{-1}, we have $h_2^{-1}h_1 = k_2k_1^{-1}$. Now the left-hand side belongs to H while the right-hand side belongs to K. Since $H \cap K = \{e\}$, it follows that $h_2^{-1}h_1 = k_2k_1^{-1} = e$ from which we conclude that $h_1 = h_2$ and $k_1 = k_2$. Hence f is one–one.

We will now prove f is a homomorphism. Let $h, h_1 \in H$ and $k, k_1 \in K$. Then $f((h, k)(h_1, k_1)) = f((hh_1, kk_1)) = hh_1kk_1$ (by definition of f). Now consider $h_1kh_1^{-1}k^{-1}$. Since $h_1 \in H$ and $kh_1^{-1}k^{-1} \in H$ (since H is a normal subgroup), it follows that $h_1kh_1^{-1}k^{-1} \in H$. Again $h_1kh_1^{-1} \in K$ (since K is a normal subgroup) and $k^{-1} \in K$. Hence $h_1kh_1^{-1}k^{-1} \in K$. Thus $h_1kh_1^{-1}k^{-1} \in H \cap K = \{e\}$ from which we can conclude that $h_1k = kh_1$. Hence $f((h, k)(h_1, k_1)) = hh_1kk_1 = hkh_1k_1 = f((h, k))f((h_1, k_1))$ (by definition of f). This proves that f is a homomorphism.

Exercises

1. Let G be a group. Show that the function $f : G \to G$ defined by $f(a) = a^{-1}$ is an isomorphism if and only if G is abelien.
2. Let G be a group. Show that the function $f : G \to G$ defined by $f(a) = a^2$ is a homomorphism if and only if G is abelian.
3. Let G be a group and a be a fixed element of G. Show that the function $f_a : G \to G$ defined by $f_a(x) = axa^{-1}$ is an isomorphism.
4. Find all normal subgroups of S_3.
5. If H and K are normal subgroups of G, prove that $H \cap K$ is also a normal subgroup of G.
6. Let G_1 and G_2 be groups and let $f : G_1 \times G_2 \to G_2$ be defined as $f((g_1, g_2)) = g_2$. Prove that f is a homomorphism. Also compute $\ker(f)$.

7. Let f be a homomorphism from a group G_1 onto an abelian group G_2. Show that $\ker(f)$ contains all elements of G_1 of the form $aba^{-1}b^{-1}$, where a and b are arbitrary elements of G_1.
8. Let $(G, *)$ be a group of even order. Let $(H, *)$ be a subgroup of $(G, *)$ where $|H| = \frac{|G|}{2}$. Show that $(H, *)$ is a normal subgroup.
9. Let H and N be subgroups of a group G. Prove that if N is a normal subgroup of G, then $H \cap N$ is a normal subgroup of H.

11 More on Algebraic Structures

In Chapter 10 we considered algebraic structures which have only one binary operation. We have seen that $(Z, +)$ is a commutative group where Z denotes the set of all integers and $+$ denotes addition operation. In fact, $\times$ (multiplication) is also a binary operation on Z. (If we multiply two integers, we get an integer.) We know that $\times$ is associative. For any three integers m, n and p, we have $m \times (n \times p) = (m \times n) \times p$. We will see how $\times$ is related to $+$.

Again, for any three integers m, n and p, we have

$$m \times (n + p) = (m \times n) + (m \times p) \text{ and}$$
$$(m + n) \times p = (m \times p) + (n \times p).$$

We say $\times$ is distributive with respect to $+$. $(Z, +, \times)$ is an example of a ring.

Definition A non-empty set R with two binary operations $+$ and $*$ is said to be a ring if the following conditions are satisfied.

1. $(R, +)$ is a commutative group.
2. $*$ is associative.
3. $*$ is distributive with respect to $+$. In other words, if $a, b, c \in R$, then

$$a * (b + c) = (a * b) + (a * c) \text{ and } (a + b) * c = (a * c) + (b * c).$$

We will look at all the properties which $\times$ on Z satisfies. We know $\times$ is commutative. For any two integers m and n, we have $m \times n = n \times m$. We say $(Z, +, \times)$ is a commutative ring.

Definition A ring $(R, +, *)$ is said to be commutative if $*$ is commutative. In other words, if $x, y \in R$, then $x * y = y * x$.

We also know that the integer 1 has the property that $1 \times n = n \times 1 = n$ for all $n \in Z$. In other words, 1 acts as the identity element of $(Z, \times)$. We say that $(Z, +, \times)$ is a ring with identity.

Definition A ring $(R, +, *)$ is said to be a ring with identity if there is an identity element in R with respect to $*$.

Note

1. A group is an algebraic structure with one binary operation whereas a ring is an algebraic structure with two binary operations.
2. In the ring $(R, +, *)$, $+$ and $*$ are only notations. Though they are termed addition and multiplication operations, they should not be confused with the usual addition and multiplication operations.
3. The identity element of $(R, +)$ (we know it exists and is unique since $(R, +)$ is a group) is denoted by 0. Again, it is only a notation. It should not be confused with the usual number 0. We thus have $a+0 = 0+a = a$ for all $a \in R$.
4. Since $(R, +)$ is a group, we know every element $a \in R$ has an inverse which is unique. This unique inverse of a is denoted by $-a$. We thus have $a + (-a) = (-a) + a = 0$.
5. If $(R, +, *)$ is a ring with identity element, then we can show that $(R, *)$ has a unique identity element which will be denoted by 1. Again, 1 should not be confused with the usual number 1. We thus have, $a * 1 = 1 * a = a$ for all $a \in R$.

Examples

1. $(Q, +, \times)$, $(\mathbb{R}, +, \times)$ and $(C, +, \times)$ are all commutative rings with identity elements where $+$ denotes usual addition and $\times$ denotes usual multiplication operations.
2. Consider $Z_6 = \{0, 1, 2, 3, 4, 5\}$. Define $\oplus$ and $\otimes$ on Z_6 as follows.

⊕	0	1	2	3	4	5
0	0	1	2	3	4	5
1	1	2	3	4	5	0
2	2	3	4	5	0	1
3	3	4	5	0	1	2
4	4	5	0	1	2	3
5	5	0	1	2	3	4

⊗	0	1	2	3	4	5
0	0	0	0	0	0	0
1	0	1	2	3	4	5
2	0	2	4	0	2	4
3	0	3	1	3	0	3
4	0	4	2	0	4	2
5	0	5	4	3	2	1

For adding (note that we are using the word adding in a loose sense) two elements of Z_6, we first add them as usual integers, then divide the sum by 6 and take the remainder. Since the remainder obtained when the sum is divided by 6 lies between 0 and 5, we are assured that the resulting element also belongs to Z_6. Similarly, for $\otimes$. Here we multiply and divide the product by 6. We can see that $(Z_6, \oplus, \otimes)$ is a commutative ring with identity element. The identity element of $(Z_6, \oplus)$ is 0, inverse of 1 with respect to $\oplus$ is 5, inverse of 2 with respect to $\oplus$ is 4, inverse of 3 with respect to $\oplus$ is 3, inverse of 4 with respect to $\oplus$ is 2 and inverse of 5 with respect to $\oplus$ is 1. The identity element of $(Z_6, \otimes)$ is 1.

3. Consider $Z_7 = \{0, 1, 2, 3, 4, 5, 6\}$. Define $\oplus$ and $\otimes$ on Z_7 as follows.

$\oplus$	0	1	2	3	4	5	6
0	0	1	2	3	4	5	6
1	1	2	3	4	5	6	0
2	2	3	4	5	6	0	1
3	3	4	5	6	0	1	2
4	4	5	6	0	1	2	3
5	5	6	0	1	2	3	4
6	6	0	1	2	3	4	5

$\otimes$	0	1	2	3	4	5	6
0	0	0	0	0	0	0	0
1	0	1	2	3	4	5	6
2	0	2	4	6	1	3	5
3	0	3	6	2	5	1	4
4	0	4	1	5	2	6	3
5	0	5	3	1	6	4	2
6	0	6	5	4	3	2	1

$\oplus$ and $\otimes$ are defined exactly as in the case of Z_6. Instead of dividing by 6, here we divide by 7. We can see that $(Z_7, \oplus, \otimes)$ is a commutative ring with identity element. The identity element of $(Z_7, \oplus)$ is 0. Inverse of 1 with respect to $\oplus$ is 6, inverse of 2 with respect to $\oplus$ is 5, inverse of 3 with respect to $\oplus$ is 4, inverse of 4 with respect to $\oplus$ is 3, inverse of 5 with respect to $\oplus$ is 2 and inverse of 6 with respect to $\oplus$ is 1. The identity element of $(Z_7, \otimes)$ is 1.

4. Generalising Examples 2 and 3, we obtain that for any positive integer n, $(Z_n, \oplus, \otimes)$ is a commutative ring with identity element. Once again, for adding two elements of Z_n, we first add them as usual integers,

then divide the sum by n and take the remainder. Since the remainder obtained when the sum is divided by n lies between 0 and $n-1$, we are assured that the resulting element also belongs to Z_n. For multiplying two elements of Z_n, we first multiply them as usual integers, then divide the product by n and take the remainder. The identity element of $(Z_n, \oplus)$ is 0 and the identity element of $(Z_n, \otimes)$ is 1.

Theorem In a ring $(R, +, *)$, the following are true.

(i) For any element $a \in R, a * 0 = 0 * a = 0$.

(ii) For any two elements $a, b \in R, (-a) * b = a * (-b) = -(a * b)$.

(iii) For any three elements $a, b, c \in R, a * (b - c) = a * b - a * c$ and $(a - b) * c = a * c - b * c$.

Proof

(i) Using the facts that 0 is the identity element of $(R, +)$ and distributivity, we obtain $0 + a * 0 = a * 0 = a * (0 + 0) = a * 0 + a * 0$. Using the right-cancellation property which holds since $(R, +)$ is a group, we can cancel $a * 0$ from both sides of $0 + a * 0 = a * 0 + a * 0$ and obtain $a * 0 = 0$. Exactly on the same lines, we can prove that $0 * a = 0$.

(ii) To prove that $(-a) * b = -(a * b)$, we have to prove that $(-a) * b$ is the inverse of $a * b$ in $(R, +)$. Again by distributivity we obtain

$$(-a) * b + a * b = (-a + a) * b = 0 * b = 0 \qquad \text{(by (}i\text{))}.$$

This proves that

$$(-a) * b = -(a * b).$$

Exactly on the same lines, we can prove that $a * (-b) = -(a * b)$.

(iii) We have

$$\begin{aligned} a * (b - c) &= a * (b + (-c)) = a * b + a * (-c) \\ &= a * b - a * c \qquad \text{(by (}ii\text{))}. \end{aligned}$$

Similarly, we can prove that $(a - b) * c = a * c - b * c$.

We observe that in $(Z_6, \oplus, \otimes)$, there are several pairs of elements (for example, 2 and 3) none of which is zero but their product (remember we are using the word product in a loose sense) is zero. We know this cannot happen in Z. (If the product of two integers is zero, then at least one of them has to be zero). Look at the table of $\otimes$ in Z_7. We can see that if $x \otimes y = 0$, then either $x = 0$ or $y = 0$.

$(Z, +, \times)$ and $(Z_7, \oplus, \otimes)$ are examples of integral domains whereas $(Z_6, \oplus, \otimes)$ is not an integral domain.

Definition In a ring $(R, +, *)$, elements x and y are said to be zero divisors if $x \neq 0$, $y \neq 0$ but $x * y = 0$.

Definition A commutative ring $(R, +, *)$ with identity is said to be an integral domain if R has no zero divisors. In other words, in an integral domain $(R, +, *)$, whenever $a * b = 0$, either $a = 0$ or $b = 0$.

Example $(Q, +, \times)$, $(\mathbb{R}, +, \times)$ and $(C, +, \times)$ are all integral domains.

Theorem $(Z_n, \oplus, \otimes)$ is an integral domain if and only of n is prime.

Proof First assuming that $(Z_n, \oplus, \otimes)$ is an integral domain, we will prove that n is prime. If n is not prime, then $n = ab(1 < a < n$ and $1 < b < n)$. Here ab denotes the usual product of the integers a and b. When we divide ab by n, the remainder is 0, proving that $a \otimes b = 0$, where $a \neq 0$ and $b \neq 0$ (since a $< n$ and $b < n$). This contradicts the fact that $(Z_n, \oplus, \otimes)$ is an integral domain.

Conversely, assume that n is prime. If $a \otimes b = 0$, then by the definition of $\otimes$, it follows that ab (product of a and b as integers) is a multiple of n so that n divides ab. Since n is a prime number, either n divides a or n divides b. If n divides a, then $a = 0$ in $(Z_n, \oplus, \otimes)$. If n divides b, then $b = 0$ in $(Z_n, \oplus, \otimes)$. This proves that $(Z_n, \oplus, \otimes)$ is an integral domain.

Now look at the table for $\otimes$ in $(Z_6, \oplus, \otimes)$. We see that there are elements (for example the element 2) which do not have inverses. Note that there is no 1 in the row corresponding to 2. Also, the elements 2 and 4 in this row are repeated. We know this cannot happen in a group. Thus $(Z_6, \otimes)$ is not a group.

Now look at the table for $\otimes$ in $(Z_7, \oplus, \otimes)$. We see that the inverse of 1 is 1, the inverse of 2 is 4, the inverse of 3 is 5, the inverse of 4 is 2, the inverse of 5 is 3 and the inverse of 6 is 6 itself. Thus every element of $(Z_7, \otimes)$ other than 0 has an inverse proving that $(Z_7 - \{0\}, \otimes)$ is a group. $(Z_7, \oplus, \otimes)$ is a field whereas $(Z_6, \oplus, \otimes)$ is not a field.

Definition A non-empty set F with two binary operations $+$ and $*$ is said to be a field if the following conditions are satisfied.

(i) $(F, +)$ is a commutative group.
(ii) $(F - \{0\}, *)$ is a commutative group.
(iii) $*$ is distributive with respect to $+$. In other words, if $x, y, z \in F$, then $x * (y + z) = x * y + x * z$ and $(x + y) * z = x * z + y * z$.

Note Again, 0 denotes the identity element of $(F, +)$, $-x$ denotes the inverse of x in $(F, +)$ and 1 denotes the identity element of $(F - \{0\}, *)$. Since $(F - \{0\}, *)$ is a group, every element $x \in F, x \neq 0$ has an inverse which is unique. This unique inverse of x is denoted by x^{-1}.

Examples

(i) $(Z, +, \times)$ is not a field. We know there are many integers (for example 4) which do not have inverses with respect to multiplication. This is an example of an integral domain which is not a field.

(ii) $(Q, +, \times)$, $(\mathbb{R}, +, \times)$ and $(C, +, \times)$ are all fields.

We see from example (i) that there are integral domains which are not fields. However, following two theorems are true.

Theorem Every field is an integral domain.

Proof Let $(F, +, *)$ be a field. To prove that $(F, +, *)$ is an integral domain, we have to prove that it has no zero divisors. This means that whenever $x * y = 0$, we have to prove that either $x = 0$ or $y = 0$. It suffices to prove that $y = 0$ assuming that $x \neq 0$. Since $x \neq 0$, $x \in F - \{0\}$. Since $(F - \{0\}, *)$ is a group, x has an inverse x^{-1}. Multiplying (applying $*$ operation) both sides of $x * y = 0$ by x^{-1}, on the left side we obtain $x^{-1} * (x * y) = x^{-1} * 0 = 0$. By associativity, we obtain $(x^{-1} * x) * y = 0$ so that $1 * y = 0$ $(x^{-1} * x = 1$ since x^{-1} is the inverse of $x)$. Since 1 is the identity element of $(F - \{0\}, *)$, it follows that $y = 1 * y = 0$.

Theorem A finite integral domain is a field.

Proof Let $(R, +, *)$ be a finite integral domain. Then $(R, +)$ is a commutative group, $*$ is both commutative and associative, there is an identity element with respect to $*$, and further $*$ is distributive with respect to $+$. The only thing that remains to be proved is that every element $x \in R - \{0\}$ has an inverse with respect to $*$ which we will now proceed to prove.

Since R is finite and R contains 0, we can write R as $R = \{0, x_1, x_2, \ldots, x_n\}$. Note that one of the $x_i s$ has to be 1. Consider an element $x \in R, x \neq 0$. We will prove that x has an inverse. Consider the elements $x * x_1, x * x_2, \ldots, x * x_n$. First, we note that none of these elements is 0. If $x * x_k = 0$ for some $k (1 \leq k \leq n)$, then R being an integral domain will mean that either $x = 0$ or $x_k = 0$, both of which are not true. We next observe that all these elements are distinct. If $x * x_i = x * x_j (i \neq j)$, then $x * (x_i - x_j) = x * x_i - x * x_j = 0$. Since R is an integral domain and $x \neq 0$, it follows that $x_i - x_j = 0$ so that $x_i = x_j (i \neq j)$ which is not true. Hence $x * x_1, x * x_2, \ldots, x * x_n$ constitute n elements of R other than 0. Since R has only $n + 1$ elements (including 0) and R contains 1, $x * x_j = 1$ for some j. Since R is commutative, it follows that $x * x_j = x_j * x = 1$, proving that x_j is the inverse of x.

Solved problems

1. Let $(A, *_1, *_2)$ be an algebraic system in which $a *_1 b = a$ for all a, b in A and let $*_2$ be an arbitrary binary operation. Prove that $*_2$ is distributive over $*_1$.

Solution For any three elements a, b and c, we have

$$a *_2 (b *_1 c) = a *_2 b \text{ (since } b *_1 c = b \text{ by our assumption).}$$

$$(a *_2 b) *_1 (a *_2 c) = a *_2 b \text{ (again by assumption).}$$

Again, $(a *_1 b) *_2 c = a *_2 c$ (since $a *_1 b = a$ by our assumption).

$$(a *_2 c) *_1 (b *_2 c) = a *_2 c \text{ (again by assumption).}$$

We have thus proved that $a *_2 (b *_1 c) = (a *_2 b) *_1 (a *_2 c)$ and $(a *_1 b) *_2 c = (a *_2 c) *_1 (b *_2 c)$, proving that $*_2$ is distributive over $*_1$.

2. Let $(R, +, *)$ be a ring such that $a * a = a$ for all $a \in R$. Prove the following.

(i) $a + a = 0$ for all $a \in R$.
(ii) $*$ is commutative.

Solution

(i) Since $*$ is distributive with respect to $+$, we have

$$(a + a) * (a + a) = a * a + a * a + a * a + a * a.$$

Since $x * x = x$ for all x, we obtain $a + a = a + a + a + a$. Since 0 is the identity element of $(R, +)$, we have

$$a + 0 + 0 + a = a + a + a + a.$$

Since $(R, +)$ is a group, left-cancellation and right-cancellation properties hold. Cancelling a on the left side and a on the right side, we obtain $a + a = 0 + 0 = 0$.

(ii) Using the assumption $a * a = a$ for all a, we obtain for all a, b,

$$\begin{aligned} a + b &= (a + b) * (a + b) \\ &= a * a + a * b + b * a + b * b \text{ (by distributivity)} \\ &= a + a * b + b * a + b \text{ (since } a * a = a \\ &\qquad \text{and } b * b = b \text{ by assumption).} \end{aligned}$$

Since 0 is the identity element of $(R, +)$, we have

$$a + 0 + 0 + b = a + a * b + b * a + b.$$

Since $(R, +)$ is a group, left-cancellation and right-cancellation properties hold. Cancelling a on the left side and b on the right side, we obtain $a * b + b * a = 0 + 0 = 0 = a * b + a * b$ (using $a + a = 0$ for all a, which is proved in (i)). Again, cancelling $a * b$ on the left side, we obtain $a * b = b * a$, proving that $*$ is commutative.

3. Let $(A, +, *)$ be an algebraic system in which e_1 is the identity element with respect to $+$ and e_2 is the identity element with respect to $*$. Further, assume that $*$ is distributive with respect to $+$ and $+$ is distributive with respect to $*$. Prove that $x + x = x$ and $x * x = x$ for all $x \in A$.

Solution Since $*$ is distributive with respect to $+$, we have

$$a * (b + c) = a * b + a * c \text{ and } (a + b) * c = a * c - b * c \quad \text{for all } a, b, c \in A. \tag{3.1}$$

Since $+$ is distributive with respect to $*$, we have

$$a + (b * c) = (a + b) * (a + c)$$

$$\text{and} \quad (a * b) + c = (a + c) * (b + c) \text{ for all } a, b, c \in A. \tag{3.2}$$

We will first prove that $e_2 + e_2 = e_2$ and $e_1 * e_1 = e_1$. We have

$$\begin{aligned} e_2 + (e_1 * e_2) &= (e_2 + e_1) * (e_2 + e_2) && \text{(by (3.2))} \\ &= e_2 * (e_2 + e_2) && \text{(since } e_1 \text{ is the identity element with respect to } +) \\ &= e_2 * e_2 + e_2 * e_2 && \text{(by (3.1))} \\ &= e_2 + e_2 && \text{(since } e_2 \text{ is the identity element with respect to } *). \end{aligned}$$

Since e_2 is the identity element with respect to $*$ and ϵ_1 is the identity element with respect to $+$, it follows that $e_2 = e_2 + e_1 = e_2 + (e_1 * e_2) = e_2 + e_2$ (by what we have proved above).

$$\begin{aligned} \text{Again, } e_1 * (e_2 + e_1) &= e_1 * e_2 + (e_1 * e_1) && \text{(by (3.1))} \\ &= e_1 + (e_1 * e_1) && \text{(since } e_2 \text{ is the identity element with respect to } *). \\ &= (e_1 + e_1) * (e_1 + e_1) && \text{(by (3.2))} \\ &= e_1 * e_1 && \text{(since } e_1 \text{ is the identity element with respect to } +). \end{aligned}$$

Since e_1 is the identity element with respect to + and e_2 is the identity element with respect to $*$, it follows that

$$e_1 = e_1 * e_2 = e_1 * (e_2 + e_1) = e_1 * e_1 \text{ (by what we have proved above).}$$

Now for every $x \in A$, we have

$$\begin{aligned} x &= x * e_2 \text{ (since } e_2 \text{ is the identity element with respect to } * \text{).} \\ &= x * (e_2 + e_2) \text{ (by what we have proved).} \\ &= x * e_2 + x * e_2 \text{ (by distributivity).} \\ &= x + x \text{ (since } e_2 \text{ is the identity element with respect to } * \text{).} \end{aligned}$$

Again, for every $x \in A$, we have

$$\begin{aligned} x &= x + e_1 \text{ (since } e_1 \text{ is the identity element with respect to } + \text{).} \\ &= x + (e_1 * e_1) \text{ (by what we have proved).} \\ &= (x + e_1) * (x + e_1) \text{ (by distributivity).} \\ &= x * x \text{ (since } e_1 \text{ is the identity element with respect to } + \text{).} \end{aligned}$$

12 More on Recurrence Relations

We have already introduced recurrence relations in Chapter 5. Recurrence relations play a vital role in combinatorics and other applications. We will see in later chapters how the concept of recurrence relations is judiciously used for solving problems of day to day interest.

Consider the following problem.

Determine the number of regions into which n lines in a plane in general positions (no two of which are parallel and no three of which are concurrent) divide the plane.

Solution Before considering n lines, we will look at $n-1$ lines. Assume that a_{n-1} is the number of regions into which these $n-1$ lines divide the plane. We will now consider adding an nth line. Since the lines are in general positions, this line has to meet all the other $n-1$ lines and the points of intersection of the nth line with all the other $n-1$ lines will have to be distinct. (Remember that no three lines are concurrent.) Assume the points of intersection to be $A_1, A_2, \ldots, A_{n-1}$. Refer to the figure given below.

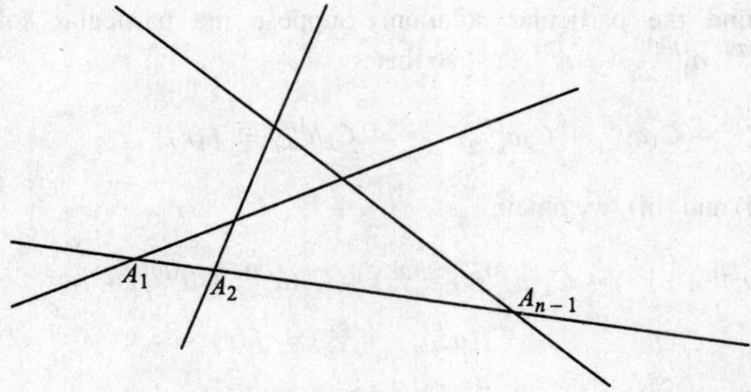

The plane already has a_{n-1} regions and now the segments $A_1A_2, A_2A_3, \ldots, A_{n-2}A_{n-1}$ have each added one extra region. The portion of the line from one end of the line to A_1 and the portion of the line from the other end of the line to A_{n-1} each contribute one extra region. Hence the total number of regions that are added to the already existing a_{n-1} regions is $n - 2 + 1 + 1 = n$. (Note that $n - 2$ comes from $A_1A_2, A_2A_3, \ldots, A_{n-2}A_{n-1}$.)

This gives the recurrence relation $a_n = a_{n-1} + n$. If we solve this recurrence relation we obtain a_n, which gives the number of regions into which n lines divide the plane. Thus the problem reduces to solving the above recurrence relation which we did in Chapter 5 and got the solution as $a_n = \frac{n(n+1)}{2}$.

In Chapter 5 we considered solving only the simplest possible recurrence relations which were linear and homogenous. A recurrence relation of the form

$$C_0 a_r + C_1 a_{r-1} + C_2 a_{r-2} + \cdots + C_k a_{r-k} = f(r) \tag{i}$$

is said to be a recurrence relation with constant coefficients if $C_0, C_1, \ldots, C_k$ are constants. The recurrence relation is said to be a kth order recurrence relation if C_0 and C_k are not zeros. Note that in the recurrence relations we will be considering in this chapter, $f(r)$ is not zero unlike the recurrence relations we considered in Chapter 5.

To solve the above recurrence relation, we will first replace $f(r)$ with 0 and solve the corresponding homogenous recurrence relation. We know how this is done from Chapter 5. Suppose $a^{(h)} = (a_0^{(h)}, a_1^{(h)}, \ldots, a_r^{(h)}, \ldots)$ denotes the homogenous solution so that

$$C_0 a_r^{(h)} + C_1 a_{r-1}^{(h)} + C_2 a_{r-2}^{(h)} + \cdots + C_k a_{r-k}^{(h)} = 0. \tag{ii}$$

We will then find the particular solution to (i). Later on, we will see how to find the particular solution. Suppose the particular solution is $a^{(p)} = (a_0^{(p)}, a_1^{(p)}, \ldots, a_r^{(p)}, \ldots)$ so that

$$C_0 a_r^{(p)} + C_1 a_{r-1}^{(p)} + C_2 a_{r-2}^{(p)} + \cdots + C_k a_{r-k}^{(p)} = f(r). \tag{iii}$$

Adding (ii) and (iii), we obtain

$$C_0(a_r^{(h)} + a_r^{(p)}) + C_1(a_{r-1}^{(h)} + a_{r-1}^{(p)}) + C_2(a_{r-2}^{(h)} + a_{r-2}^{(p)}) + \cdots + C_k(a_{r-k}^{(h)} + a_{r-k}^{(p)}) = f(r).$$

This proves that the total solution $a^{(h)} + a^{(p)}$ also satisfies the recurrence relation (i).

We will illustrate the method of finding the particular solution and hence the total solution with the help of some examples.

Example 1 Solve the recurrence relation $a_r - 7a_{r-1} + 10a_{r-2} = 3^r$ given that $a_0 = 0$ and $a_1 = 1$.

Solution Since this is a second order recurrence relation with constant coefficients, we need two initial conditions to determine the two constants. The corresponding homogenous recurrence relation is obtained by replacing 3^r on the right-hand side with 0. In this case it is $a_r - 7a_{r-1} + 10a_{r-2} = 0$. The characteristic equation is $x^2 - 7x + 10 = 0$ whose roots are 5 and 2. Since the roots of the characteristic equation are real and distinct, the homogeneous solution is given by

$$a_r^{(h)} = k_1 5^r + k_2 2^r.$$

Since the right-hand side is of the form 3^r, we will take the particular solution to be $P \times 3^r$. Substituting this in the given recurrence relation, we obtain

$$P \times 3^r - 7P \times 3^{r-1} + 10P \times 3^{r-2} = 3^r.$$

This is the same as

$$P \times 3^r - \frac{7}{3}P \times 3^r + \frac{10}{9}P \times 3^r = 3^r,$$

which is the same as

$$P \times 3^r \times \left(1 - \frac{7}{3} + \frac{10}{9}\right) = 3^r.$$

Hence, $P = \frac{-9}{2}$ so that the total solution is

$$a_r = k_1 5^r + k_2 2^r - \frac{9}{2} \times 3^r.$$

Using the initial conditions $a_0 = 0$ and $a_1 = 1$, we obtain $k_1 + k_2 = \frac{9}{2}$ and $5k_1 + 2k_2 = \frac{29}{2}$. Solving, we obtain $k_1 = \frac{11}{6}$ and $k_2 = \frac{8}{3}$. Hence the complete solution to the given recurrence relation is given by

$$a_r = \frac{11}{6} \times 5^r + \frac{8}{3} \times 2^r - \frac{9}{2} \times 3^r.$$

Example 2 Solve the recurrence relation $a_r + 6a_{r-1} + 9a_{r-2} = 3$ given that $a_0 = -3$ and $a_1 = 4$.

Solution The characteristic equation is $x^2 + 6x + 9 = 0$ whose roots are -3 and -3. Since the roots are real and equal, the homogeneous solution is given by

$$a_r^{(h)} = k_1(-3)^r + k_2 \times r \times (-3)^r.$$

Since the right-hand side is a constant, the particular solution is also a constant, say P. Substituting this in the given recurrence relation we obtain $P + 6P + 9P = 3$. This is the same as $16P = 3$ from which we obtain $P = \frac{3}{16}$. Hence the total solution is

$$a_r = k_1(-3)^r + k_2 \times r \times (-3)^r + \frac{3}{16}.$$

Using the initial conditions $a_0 = -3$ and $a_1 = 4$, we obtain $k_1 = \frac{-51}{16}$ and $k_2 = \frac{23}{12}$.

Hence the complete solution to the given recurrence relation is given by

$$a_r = \frac{-51}{16}(-3)^r + \frac{23}{12} \times r \times (-3)^r + \frac{3}{16}.$$

12.1 Numeric Functions and Generating Functions

A numeric function is simply a function whose domain is the set of all positive integers and whose range is the set of all real numbers. In other words, a numeric function is a function $f : Z^+ \to \mathbb{R}$ where Z^+ denotes the set of all positive integers and $\mathbb{R}$ denotes the set of all real numbers. $f(0), f(1), f(2), \ldots$ are denoted by $a_0, a_1, a_2, \ldots$ respectively. In general, $f(i)$ is denoted by a_i.

With a numeric function $a = (a_0, a_1, a_2, \ldots, a_i, \ldots)$, we can associate a formal power series $a_0 + a_1 z + a_2 z^2 + \cdots + a_i z^i + \cdots$ which is called the generating function associated with the numeric function a.

Example Consider the numeric function $(5^0, 5^1, 5^2, \ldots, 5^i, \ldots)$. The corresponding generating function is $5^0 + 5^1 z + 5^2 z^2 + \cdots + 5^i z^i + \cdots$, which is nothing but $\frac{1}{(1-5z)}$.

If the generating function corresponding to a is $A(z)$, then the generating function corresponding to ca is $cA(z)$.

Example The generating function corresponding to the numeric function $c_r = 7^{r+3} (r \geq 0)$ is $C(z) = \frac{343}{(1-7z)}$ since $7^3 = 343$.

We will now illustrate how generating functions are used in solving recurrence relations. Consider the problem given at the beginning of this chapter. We have seen that the corresponding recurrence relation is $a_n = a_{n-1} + n$. We will now solve this recurrence relation using the method of generating functions. We have

$$a_n - a_{n-1} = n(n \geq 1).$$

Multiplying both sides by t^n, we obtain

$$a_n t^n - a_{n-1} t^n = n t^n,$$

$$\sum_{k=1}^{\infty} a_k t^k - \sum_{k=1}^{\infty} a_{k-1} t^k = \sum_{k=1}^{\infty} k t^k.$$

Let $f(t) = \sum_{k=0}^{\infty} a_k t^k$ where $a_0 = 0$. Then,

$$f(t) - t f(t) = \sum_{k=0}^{\infty} k t^k$$

$$f(t) = \sum_{k=0}^{\infty} \frac{k t^k}{(1-t)} = \sum_{k=1}^{\infty} \frac{k t^k}{(1-t)}. \quad \text{(i)}$$

We have $\quad \dfrac{1}{(1-t)} = \sum_{k=0}^{\infty} t^k.$

Application of the D-operator gives

$$\frac{1}{(1-t)^2} = \sum_{k=1}^{\infty} k t^{k-1}.$$

Multiplying both sides by t, we obtain

$$\frac{t}{(1-t)^2} = \sum_{k=1}^{\infty} k t^k.$$

Substituting this in (i), we obtain

$$f(t) = \frac{t}{(1-t)^3} = t(1-t)^{-3}$$

$$= t\left(1 + 3t + \frac{(3 \times 4)}{(1 \times 2)} t^2 + \cdots \frac{k(k+1)}{(1 \times 2)} t^{k-1} + \cdots\right)$$

$$= \left(t + 3t^2 + \frac{(3 \times 4)}{(1 \times 2)} t^3 + \cdots \frac{k(k+1)}{(1 \times 2)} t^k + \cdots\right).$$

Equating the coefficient of t^k on both sides, we obtain $a_k = \frac{k(k+1)}{2}$. Hence $a_n = \frac{n(n+1)}{2}$ is the number of regions into which n lines divide the plane.

We will now solve a recurrence relation which is an example of a recurrence relation with variable coefficients.

$$a_n - na_{n-1} = -(a_{n-1} - (n-1)a_{n-2})$$

given that $a_0 = 1$ and $a_1 = 0$.

We have

$$a_n - na_{n-1} = -(a_{n-1} - (n-1)a_{n-2}) = -[-(a_{n-2} - (n-2)a_{n-3})].$$

We can now replace $a_{n-2} - (n-2)a_{n-3}$ by $-(a_{n-3} - (n-3)a_{n-4})$. Repeating this, we obtain

$$\begin{aligned} a_n - na_{n-1} &= (-1)^{n-1}(a_{n-(n-1)} - (n-(n-1))a_{1-1}) \\ &= (-1)^{n-1}(a_1 - a_0) = (-1)^n \end{aligned}$$

since $a_0 = 1$ and $a_1 = 0$.

Thus the recurrence relation becomes

$$a_n - na_{n-1} = (-1)^n \quad (n \geq 1).$$

We will now make use of exponential generating function. Replacing n by k and taking the summation, we obtain

$$\sum_{k=1}^{\alpha} a_k \frac{t^k}{k!} - \sum_{k=1}^{\alpha} ka_{k-1} \frac{t^k}{k!} = \sum_{k=1}^{\alpha} (-1)^k \frac{t^k}{k!}.$$

Let $f(t) = \sum\limits_{k=0}^{\alpha} a_k \frac{t^k}{k!}$.

Then $f(t) - 1 - tf(t) = e^{-t} - 1$ so that $f(t) = \frac{e^{-t}}{(1-t)}$. Using the expansion of e^{-t} and $\frac{1}{(1-t)}$, we obtain

$$f(t) = \left(1 - t + \frac{t^2}{2!} - \frac{t^3}{3!} + \cdots\right)(1 + t + t^2 + t^3 + \cdots).$$

Coefficient of t^k on the right-hand side is

$$1 - \frac{1}{1!} + \frac{1}{2!} - \frac{1}{3!} + \cdots + \frac{(-1)^k}{k!}.$$

Coefficient of t^k in $f(t)$ is $\frac{a_k}{k!}$. Equating the two, we obtain

$$a_k = k!\left(1 - \frac{1}{1!} + \frac{1}{2!} - \frac{1}{3!} + \cdots + \frac{(-1)^k}{k!}\right).$$

13 More on Functions

In this chapter our main focus will be on answering the following questions.

1. How many functions exist from a set consisting of n elements into a set consisting of m elements ?
2. How many of the functions in 1 are one–one ?
3. How many of the functions in 1 are onto ?

We will see that the answer to question 1 is straightforward whereas the answer to question 2 is little involved. Answer to question 3 will require considerable effort. For the sake of simplicity, we will refer to a set consisting of n elements as an n-set.

We know that in English language any word consists of letters chosen from A to Z (or a to z). These 26 capital letters or 26 small letters constitute the alphabets of English language. Similarly, every natural language has its own set of alphabets. If we take an alphabet $\Sigma = \{0, 1\}$, then some of the words which can be formed using the alphabet Σ are 0010, 011101001, 1011101, etc.

Let $A = \{1, 2, 3, \ldots, n\}$ be an n-set. Consider an alphabet $B = \{a_1, a_2, \ldots, a_m\}$ consisting of m letters and consider a word of length n, say $a_{i1}\, a_{i2} \ldots a_{in}$. We note that $a_{i1}, a_{i2}, \ldots, a_{in} \in B$. We will first find out the number of such words that can be formed. Since the letters in the word need not be distinct, there are m choices for the first letter, again m choices for the second letter, and finally m choices for the nth letter. It follows that the number of such words is $m \times m \times \cdots \times m$ (n times) which is nothing but m^n.

With each word $a_{i1}\, a_{i2}\, a_{i3} \cdots a_{in}$ of length n, we can asscciate a function from A to B as follows.

$$\begin{matrix} 1 & 2 & 3 & \cdots & n \\ a_{i1} & a_{i2} & a_{i3} & & a_{in} \end{matrix}$$

Note that this is a well-defined function. This function need not be one–one since it is possible that $a_{ij} = a_{ik} (j \neq k)$. This function need not be onto also since the word need not contain all the alphabets from B.

Conversely, consider the following function from A to B.

$$\begin{matrix} 1 & 2 & 3 & n \\ a_{j1} & a_{j2} & a_{j3} & a_{jn} \end{matrix}$$

From this function we obtain $a_{j1}\, a_{j2}\, a_{j3} \ldots a_{jn}$, which is a word of length n on m alphabets.

From the above discussion we see that there is a one-to-one correspondence between the set of all functions from an n-set into an m-set and the set of all words of length n on an alphabet of m letters. Since the number of elements in the latter set is m^n, it follows that the number of functions which can be defined from an n-set into an m-set is also equal to m^n.

Suppose the elements of A stand for the labels on n distinct objects and the elements of B stand for the labels on m distinct boxes. In other words, the objects have the labels $1, 2, 3, \ldots, n$ and the boxes have the labels $a_1, a_2, \ldots, a_m$. Consider the following function from A to B.

$$\begin{matrix} 1 & 2 & 3 & n \\ a_{j1} & a_{j2} & a_{j3} & a_{jn} \end{matrix}$$

We can decide to place the object with label 1 into the box with label a_{j1}, object with label 2 into the box with label a_{j2} and finally the object with label n into the box with label a_{jn}. This gives us a distribution of n distinct objects into m distinct boxes.

Conversely, suppose we decide to place object with label 1 into the box with label a_{k1}, object with label 2 into the box with label a_{k2}, and finally the object with label n into the box with label a_{kn}. Corresponding to this distribution, we obtain the following function from A to B.

$$\begin{matrix} 1 & 2 & 3 & n \\ a_{k1} & a_{k2} & a_{k3} & a_{kn} \end{matrix}$$

From the above observations we note that there is a one-to-one correspondence between the set of all functions from an n-set into an m-set and the set of distributions of n distinct objects into m distinct boxes. Since the number of elements in the former set is m^n, it follows that the number of ways in which n distinct objects can be placed in m distinct boxes is also equal to m^n. (We do not impose any restrictions here. In other words, a box may contain more than one object. Also it is not necessary that every box should contain objects. Some boxes may be empty.)

Note Here m may be less than n or m may be equal to n or m may be greater than n.

Example Consider the alphabet $B = \{A, D, E, G, N, R, V\}$. Here $m = 7$. Consider a word of length 4 say DARA from this alphabet. Corresponding to this word, we obtain the following function from $A = \{1, 2, 3, 4\}$ into B.

1	2	3	4
D	A	R	A

We note that the other elements of B namely E, G, N and V do not have pre-images. Hence this function is not onto. It is also not one–one since 2 and 4 have the same image.

We now have four objects labelled 1, 2, 3, 4. We also have seven boxes labelled A, D, E, G, N, R and V. We place object with label 1 in the box with label D, object with label 2 in the box with label A, object with label 3 in the box with label R and object with label 4 in the box with label A. We note that the boxes with labels E, G, N and V are empty whereas the box with label A contains two objects (one object with label 2 and another object with label 4). We also note that the number of such words of length 4 that can be formed using the letters in B is 7^4.

We will now consider the problem of determining the number of functions from an n-set into an m-set which are one–one.

Consider a word of length n on an alphabet of m letters but with distinct letters. This means that no two letters in this word are the same. Obviously, this is not possible if $m < n$. Consider such a word say $a_{l1}\, a_{l2}\, a_{l3} \ldots a_{ln}$. With this word we can associate the following function from A to B which is clearly one–one.

1	2	3	$\cdots$	n
a_{l1}	a_{l2}	a_{l3}	$\cdots$	a_{ln}

Conversely, consider the following function from A to B which is one–one.

1	2	3	n
a_{l1}	a_{l2}	a_{l3}	a_{ln}

Since the function is one–one, $a_{li} \neq a_{lj} (i \neq j)$. Hence $a_{l1}\, a_{l2}\, a_{l3} \ldots a_{ln}$ is a word of length n with distinct letters.

With a word of length n with distinct letters, say $a_{p1}\, a_{p2}\, a_{p3} \ldots a_{pn}$, we can associate an n-tuple $(a_{p1}, a_{p2}, \ldots, a_{pn})$. This is an n-tuple without repetitions. Conversely, given an n-tuple without repetitions, say $(a_{q1}, a_{q2}, \ldots, a_{qn})$, we obtain $a_{q1}\, a_{q2} \ldots a_{qn}$ which is a word of length n with distinct letters.

We now assume that if an object is placed in a box, then no other object can be placed in that box. In other words, a box can contain a maximum of one object. This property is called the exclusion principle (EP).

Now consider the following function from A to B which is one–one.

$$\begin{matrix} 1 & 2 & 3 & \cdots & n \\ a_{d1} & a_{d2} & a_{d3} & \cdots & a_{dn} \end{matrix}$$

We can place the object with label 1 in the box labelled a_{d1}, object with label 2 in the box labelled a_{d2}, object with label 3 in the box labelled a_{d3} and object with label n in the box labelled a_{dn}. Since $a_{d1}, a_{d2}, \ldots, a_{dn}$ are all distinct, it follows that a box contains a maximum of one object and we thus obtain a distribution of n distinct objects into m distinct boxes with EP.

Conversely, consider a distribution of n distinct objects into m distinct boxes with EP. Assume that the object labelled 1 is placed in the box labelled a_{q1}, object labelled 2 is placed in box labelled a_{q2}, and finally the object labelled n is placed in box labelled a_{qn}. Because of the property EP, $a_{q1}, a_{q2}, \ldots, a_{qn}$ are all distinct. Hence the function given below is one–one.

$$\begin{matrix} 1 & 2 & 3 & n \\ a_{q1} & a_{q2} & a_{q3} & a_{qn} \end{matrix}$$

Let $a_{t1}\, a_{t2} \ldots a_{tn}$ be a word of length n (with distinct letters) chosen from the alphabet B. We can say that we have arranged n letters (chosen from B) in a particular order, namely a_{t1} first, a_{t2} second and finally a_{tn} last (at the nth place). Thus a word of length n (with distinct letters chosen from the alphabet B) corresponds to a permutation (arrangement) of m letters taken n at a time. This means that we are not considering all the m letters of B but only n letters and arranging them.

Conversely, pick n letters from B (say $a_{s1}, a_{s2}, \ldots, a_{sn}$) and arrange them in a particular order, say a_{s1} first, a_{s2} second, and finally a_{sn} last (at the nth place). From this arrangement we get a word of length n, namely $a_{s1}\, a_{s2} \ldots a_{sn}$. Note that if we had arranged the letters in a different order (say a_{s2} first), then we would have got a word in which a_{s2} is the first letter; this word would have been different from the above word of length n (note that $a_{s1} \neq a_{s2}$).

We have thus proved the following.

There is a one–one correspondence between any two of the following sets.

1. The set of all words of length n with distinct letters chosen from an alphabet of m letters.
2. The set of all one–one functions from an n-set into an m-set.
3. The set of all n-tuples on m letters without repetition.
4. The set of distributions of n distinct objects into m distinct boxes with EP.
5. The set of permutations of m symbols taken n at a time.

We will now consider the problem of determining the number of functions from an n-set into an m-set which are one–one. According to the above observation, this number is the same as the number of elements in the set of all n-tuples on m letters without repetition. We will now prove that this number is $m(m-1)(m-2)\cdots(m-n+1)$.

Proof Let C_n denote the set of all n-tuples on m letters without repetition. To form C_{n+1} we have to add one more element to each n-tuple in C_n. Since the elements in each of the $n+1$ tuples have to be distinct, the element that is added in each n-tuple cannot be the same as any of the elements which are already in that n-tuple. In other words, the element to be added in an n-tuple has to be chosen from the remaining $m-n$ letters. Assuming that $|C_n|$ denotes the number of elements in C_n, we obtain the following recurrence relation.

$$|C_{n+1}| = (m-n)|C_n|$$

By repeatedly applying the above recurrence relation and using the fact that $|C_1| = m$ (C_1 is after all the set of all 1-tuples on m letters and the set of all 1-tuples on m letters that can be formed is only m), we obtain the following.

$$\begin{aligned}|C_{n+1}| &= (m-n)|C_n| \\ &= (m-n)(m-n+1)|C_{n-1}| \\ &= (m-n)(m-n+1)\cdots(m-1)|C_1| \\ &= (m-n)(m-n+1)\cdots(m-1)m\end{aligned}$$

Replacing n by $n-1$ in the above, we obtain $|C_n| = m(m-1)(m-2)\cdots(m-n+1)$.

We have thus proved the following theorem.

Theorem The number of functions from an n-set A into an m-set B which are one–one is $m(m-1)(m-2)\cdots(m-n+1)$.

Deduction Suppose $m = n$. Then we know that any function from A to B which is one–one is also onto. Hence by replacing m with n in the above theorem we obtain that the number of bijections from A to B is $n(n-1)(n-2)\cdots(n-n+1)$, which is nothing but $n!$

Example Consider once again the alphabet $B = \{\text{A, D, E, G, N, R, V}\}$ and consider the word VREAD. This is a five-letter word with distinct letters. Here $n = 5$. Hence $A = \{1, 2, 3, 4, 5\}$. The corresponding one–one function from A to B is

1	2	3	4	5
V	R	E	A	D

We have taken 5 letters from B and have arranged them in the order such that V appears first, R appears second, E appears third, A appears fourth and D appears fifth.

The corresponding 5 tuple without repetitions is (V, R, E, A, D). We place the object labelled 1 in the box labelled V, object labelled 2 in box labelled R, object labelled 3 in the box labelled E, object labelled 4 in the box labelled A and object labelled 5 in the box labelled D. The other two boxes labelled G and N are empty. No box contains more than one object and hence EP is satisfied. The number of ways in which the placement of objects into the boxes can be done is

$$7 \times 6 \times 5 \times 4 \times 3 = 2520.$$

This is also the number of five-letter words that can be formed without repetitions using the alphabet B.

We will now consider the problem of determining the number of functions from an n-set into an m-set which are onto. But this requires some amount of work.

Problem Suppose we have four distinct objects labelled 1, 2, 3, 4 which we want to place in three boxes. We assume the following.

1. Every object should go to one and only one box.
2. No box can be empty. In other words, every box should contain at least one object.
3. The boxes are identical. In other words, the boxes do not have labels. (The meaning of this will become clear in a short while.) What is the number of ways in which this distribution can be done?

Note It is possible to place the objects in boxes satisfying the above conditions only when the number of objects is greater than or equal to the number of boxes.

Consider the following alternatives.

(a) $\{1, 2\}, \{3\}, \{4\}$
(b) $\{1, 3\}, \{2\}, \{4\}$
(c) $\{1, 4\}, \{2\}, \{3\}$
(d) $\{2, 3\}, \{1\}, \{4\}$
(e) $\{2, 4\}, \{3\}, \{1\}$
(f) $\{3, 4\}, \{1\}, \{2\}$

Alternative (a) says that one way of partitioning the 4 objects in 3 boxes is place objects with labels 1 and 2 in one box (it does not matter in which box since the boxes are assumed to be identical), object with label 3 in another box and object with label 4 in yet another box. Please note that we do not consider the alternatives {3}, {4}, {1, 2} or {4}, {1, 2}, {3}. These alternatives are identical to alternative (a). This is the precise meaning of saying that the boxes are identical.

We see that with the given restrictions, the above six are the only alternatives that are possible.

Now consider the problem of placing five distinct objects labelled 1, 2, 3, 4, 5 into three identical boxes. Instead of listing all the possibilities as we have done above, we will proceed with the following approach.

First, we ignore object labelled 5 and distribute the four objects labelled 1, 2, 3, 4 into three identical boxes. Six ways of doing this are given above. We will now add the object labelled 5. Since there are only three boxes, alternative (a) leads to the following three alternatives.

{1, 2, 5}, {3}, {4}

{1, 2}, {3, 5}, {4}

{1, 2}, {3}, {4, 5}.

Similarly, each of the alternatives from (b) to (f) gives rise to three alternatives. Hence the total number of alternatives becomes $3 \times 6 = 18$. Have we exhausted all possibilities? No! What about the possibilities in which object labelled 5 alone is in a box? Since there are only three boxes, the other four objects have to be accommodated in the remaining two boxes. Thus we obtain the following alternatives.

{1, 2, 3}, {4}, {5}

{1, 2, 4}, {3}, {5}

{1, 3, 4}, {2}, {5}

{2, 3, 4}, {1}, {5}

{1, 2}, {3, 4}, {5}

{1, 3}, {2, 4}, {5}

{2, 3}, {1, 4}, {5}

This gives the number 7. (Note that this number is the same as the number of ways in which four distinct objects can be placed in two identical boxes). Hence the total number of ways is $18 + 7 = 25$.

The number of ways in which n distinct objects can be placed in m identical boxes is denoted by S_n^m. From the above discussion, we note that $S_4^3 = 6$ and $S_5^3 = 25$.

Now the question is how to find S_n^m ? We will make use of recurrence relations to answer this question.

Consider S_{n+1}^m which denotes the number of ways in which $n + 1$ distinct objects can be placed in m identical boxes. It may happen that a particular object (say i) is the only object in a box. Again, we do not have to bother which box since the boxes are assumed to be identical. In this case our problem reduces to the distribution of the remaining n distinct objects into the remaining $m - 1$ identical boxes. We know this can be done in S_n^{m-1} ways.

We now consider the second possibility that the object i is not the only object in any box. We remove i and distribute the remaining n distinct objects into m identical boxes; this can be done in S_n^m ways. Now every one of the m boxes contains at least one object and we will place i in one of the m boxes. There are m alternatives for placing the object i and hence the total number of ways in which the second possibility can be carried out is $m \times S_n^m$.

The above discussions lead to the following recurrence relation.

$$S_{n+1}^m = S_n^{m-1} + m \times S_n^m.$$

We will look at the values of S_n^m for some special values of m and n.

What is S_1^1 ? It is the number of ways in which one object can be placed in one box. Obviously this number is 1. What is S_n^n ? It is the number of ways in which n distinct objects can be placed in n identical boxes. Since no box should be empty, this number is also 1: simply place the first object into one of the boxes (it does not matter which box), second object into another box, etc.

Suppose $n < m$. What is S_n^m ? Since the number of objects is less than the number of boxes, it is not possible to place the objects in the boxes in such a way that no box is empty. Note that an object can go only to one box. We thus obtain the following.

$$S_1^1 = S_n^n = 1, S_n^m = 0 \text{ whenever } n < m.$$

We will try to find the value of S_6^4. Replacing m by 4 and n by 5 in the above recurrence relation, we obtain

$$S_6^4 = S_5^3 + 4 \times S_5^4 \qquad \text{(i)}$$

$$S_5^4 = S_4^3 + 4 \times S_4^4$$
$$= 6 + 4 \times 1 \text{ (Since } S_4^3 = 6 \text{ and } S_4^4 = 1)$$
$$= 10.$$

We know that $S_5^3 = 25$. Substituting these values of S_5^3 and S_5^4 in (i), we obtain

$$S_6^4 = 25 + 4 \times 10 = 65.$$

Thus there are 65 ways in which six distinct objects can be placed in four identical boxes.

The number $S_n^1 + S_n^2 + \cdots + S_n^m$ gives the total number of ways in which n distinct objects can be placed in m non-distinct boxes. Note that here some boxes may be empty. When $m = n$, we obtain $B_n = S_n^1 + S_n^2 + \cdots + S_n^n$ which is called the nth Bell number.

Theorem The number of functions from an n-set into an m-set which are onto is given by $m! \times S_n^m$.

Proof Before proving this theorem we will illustrate the logic of the proof with the help of an example.

Let $X = \{1, 2, 3\}$ and $A = \{a, b\}$. Here we have taken $n = 3$ and $m = 2$. Suppose we have an onto function $f: X \to A$ defined as $f(1) = a$, $f(2) = b$ and $f(3) = a$. This defines a partition $\{\{1, 3\}, \{2\}\}$ of X. Since the images of 1 and 3 are the same, we have included both in the same class.

Now consider all the possible partitions of X into two classes. We know there are three such partitions (note that $S_3^2 = 3$) namely, $\{\{1, 2\}, \{3\}\}$, $\{\{2, 3\}, \{1\}\}$ and $\{\{3, 1\}, \{2\}\}$. Consider the partition $\{\{1, 2\}, \{3\}\}$. This one partition gives rise to two possible functions from X onto A namely, the one which maps 1 and 2 to a and 3 to b, and the other one which maps 1 and 2 to b and 3 to a. Similarly, from each of the other two partitions, we get two onto functions. Hence the total number of onto functions is $2 \times 3 = 2! \times S_3^2$.

We will now come to the proof of the theorem. Let $X = \{x_1, x_2, \ldots, x_n\}$ and $A = \{a_1, a_2, \ldots, a_m\}$. Suppose $g: X \to A$ is an onto function. Then, $g^{-1}(a_1), g^{-1}(a_2), \ldots, g^{-1}(a_m)$ defines a partition of X. Note that each one of them is non-empty (since g is onto); the union of them is X and intersection of any two of them is empty. (Otherwise, g will not be a function.)

Now consider a partition $Y = \{Y_1, Y_2, \ldots, Y_m\}$ of X. There are m possible ways of defining the image of the elements in Y_1. They can be either a_1 or a_2 or a_m. Once we have defined the image of the elements in Y_1 in one of the m possible ways, there are $m - 1$ ways of defining the image of the elements in

Y_2. Finally, there will be only one way of defining the image of the elements in Y_m. It follows that corresponding to one partition of X into m classes, there exist $m \times (m-1) \times (m-2) \times \cdots \times 1 = m!$ onto functions. Since the number of partitions of X into m classes is S_n^m, the result follows.

Deduction Suppose $m = n$. Then we know that any function from A to B which is onto is also one–one. Hence by replacing m with n in the above theorem, we obtain that the number of bijections from A to B is $n! \times S_n^n$, which is nothing but $n!$ since $S_n^n = 1$.

14 Partitions

Consider the number 6. What are the different ways in which 6 can be written? 6 can be written as 6, $1 + 5$, $2 + 4$, $3 + 3$, $1 + 1 + 4$, $1 + 2 + 3$, $2 + 2 + 2$, $1 + 1 + 1 + 3$, $1 + 1 + 2 + 2$, $1 + 1 + 1 + 1 + 2$ and $1 + 1 + 1 + 1 + 1 + 1$. Are there any other ways? Since $1 + 1 + 2 + 2$ is considered, we will not be considering $2 + 1 + 2 + 1$, $2 + 2 + 1 + 1$, etc. Then we can see that whatever we have written above are the only possibilities. We say that 6 can be partitioned in 11 ways and denote it by writing $P(6) = 11$. We also observe that some partitions of 6 have only one part (for example, 6), some have two parts (e.g., $1 + 5$, $2 + 4$ and $3 + 3$), some have three parts (e.g., $1 + 1 + 4$, $1 + 2 + 3$ and $2 + 2 + 2$), some have four parts (e.g., $1 + 1 + 1 + 3$ and $1 + 1 + 2 + 2$), some have five parts (e.g., $1 + 1 + 1 + 1 + 2$) and some have six parts (e.g., $1 + 1 + 1 + 1 + 1 + 1$). We observe that no partition of 6 can have more than six parts. We say $P_6^1 = 1$, $P_6^2 = 3$, $P_6^3 = 3$, $P_6^4 = 2$, $P_6^5 = 1$, $P_6^6 = 1$ and $P_6^m = 0$ for all $m > 6$. Obviously, the sum of all these numbers is 11 which is $P(6)$.

What about the number 7? We observe that 7 can be written as 7, $1 + 6$, $2 + 5$, $3 + 4$, $1 + 1 + 5$, $1 + 2 + 4$, $1 + 3 + 3$, $2 + 2 + 3$, $1 + 1 + 1 + 4$, $1 + 1 + 2 + 3$, $1 + 2 + 2 + 2$, $1 + 1 + 1 + 1 + 3$, $1 + 1 + 1 + 2 + 2$, $1 + 1 + 1 + 1 + 1 + 2$ and $1 + 1 + 1 + 1 + 1 + 1 + 1$. We say $P(7) = 15$, $P_7^1 = 1$, $P_7^2 = 3$, $P_7^3 = 4$, $P_7^4 = 3$, $P_7^5 = 2$, $P_7^6 = 1$, $P_7^7 = 1$ and $P_7^m = 0$ for all $m > 7$. Obviously, the sum of all the numbers from P_7^1 to P_7^7 is 15 which is $P(7)$.

This chapter is devoted to the study of determining $P(n)$ for a positive integer n and P_n^m where $m \leq n$. We know $P_n^m = 0$ whenever $m > n$.

We first develop some preliminaries.

Assume that we have n identical objects (which means the objects do not have labels) and m identical boxes. We will consider the problem of placing these n objects in m boxes in such a way that no box is empty. Let c_1 denote the maximum number of objects which are placed in one box, c_2 the next maximum number of objects which are placed in another box, and finally c_m the least number of objects which are placed in the remaining box.

Since every object is placed in one box or the other and no box is empty, we obtain

$$c_1 + c_2 + \cdots + c_m = n \text{ where } c_1 \geq c_2 \geq \cdots c_m \geq 1.$$

The m-tuple $(c_1, c_2, \ldots, c_m)$ is called a partition of n. Suppose $n = 7$ and $m = 3$. We then obtain the following partitions of 7.

$$(5, 1, 1), (4, 2, 1), (3, 3, 1), (3, 2, 2).$$

If we want to distribute seven identical objects into three identical boxes, the various possibilities are:

1. Place five objects in one box (it does not matter which five objects are taken since the objects are identical; the box is also immaterial since the boxes are identical), one object in another box and the remaining one object in the remaining box.
2. Place four objects in one box, two objects in another box and one object in the remaining box.
3. Place three objects in one box, three objects in another box and one object in the remaining box.
4. Place three objects in one box, two objects in another box and two objects in the remaining box.

Note that since we have included (4, 2, 1), we do not include (2, 1, 4), (1, 4, 2), etc. The number 4 is the number of partitions of 7 into exactly three parts.

The number of partitions of n into exactly m parts is denoted by P_n^m. We have $P_7^3 = 4$.

Since 7 can be written using only one part as 7, it follows that $P_7^1 = 1$.

We have $P_7^2 = 3$ (the partitions of 7 using exactly two parts are (6, 1), (5, 2) and (4, 3)).

$P_7^3 = 4$ (the partitions of 7 using exactly three parts are (1, 1, 5), (1, 2, 4), (1, 3, 3) and (2, 2, 3)).

$P_7^4 = 3$ (the partitions of 7 using exactly four parts are (4, 1, 1, 1), (2, 2, 2, 1) and (3, 2, 1, 1)).

$P_7^5 = 2$ (the partitions of 7 using exactly five parts are (3, 1, 1, 1, 1), (2, 2, 1, 1, 1)).

$P_7^6 = 1$ (the only partition of 7 using exactly six parts is (2, 1, 1, 1, 1, 1)).

$P_7^7 = 1$ (the only partition of 7 using exactly seven parts is (1, 1, 1, 1, 1, 1, 1)).

The number $P_n^1 + P_n^2 + \cdots + P_n^m$ denotes the number of ways in which n can be partitioned into maximum m parts. This is also the number of ways in which n identical objects can be placed in m identical boxes (some boxes may be empty).

Examples

1. The number of partitions of 7 into four or less than four parts is

$$P_7^1 + P_7^2 + P_7^3 + P_7^4 = 11.$$

If $m = n$, then the number $P_n^1 + P_n^2 + \cdots + P_n^n$ is denoted by $P(n)$. It is the number of ways in which the integer n can be partitioned.

2. $P(7) = P_7^1 + P_7^2 + P_7^3 + P_7^4 + P_7^5 + P_7^6 + P_7^7 = 15.$

Suppose $n < m$. Then it is not possible to distribute n identical objects into m identical boxes such that no box is empty. Hence it follows that $P_n^m = 0$ whenever $n < m$. Now P_n^1 denotes the number of ways in which n can be written using only one part. There is only one way namely n. Hence $P_n^1 = 1$. Again P_n^n denotes the number of ways in which n can be written using n parts. There is only one way namely $(1, 1, 1, \ldots, 1)$ (n times). Hence $P_n^n = 1$.

We will now see how to find the values of P_n^m. Again we will make use of recurrence relation.

Let X denote the set of all partitions of n into m or less than m parts. Any element of X can be expressed as an m-tuple $(x_1, x_2, \ldots, x_m)$, where $x_1 + x_2 + \cdots + x_m = n$. Note that some of the x_is may be 0s. We have

$$|X| = P_n^1 + P_n^2 + \cdots + P_n^m.$$

Let Y denote the set of all partitions of $n + m$ into exactly m parts. We know $|Y| = P_{n+m}^m$.

Define a function f on X as follows.

$$f(x_1, x_2, \ldots, x_m) = (x_1 + 1, x_2 + 1, \ldots, x_m + 1).$$

We have

$$x_1 + 1 + x_2 + 1 + \cdots + x_m + 1 = x_1 + x_2 + \cdots + x_m + m = n + m.$$

The right-hand side is a partition of $n + m$ into exactly m parts and hence the right-hand side belongs to Y. f is thus a well-defined function from X to Y. Further, f is both one–one and onto. Hence $|X| = |Y|$, proving that

$$P_n^1 + P_n^2 + \cdots + P_n^m = P_{n+m}^m.$$

For every positive integer n, we also have $P_n^1 = 1$ and $P_n^n = 1$. In how many ways can we represent n using only one part? Only one way, namely n itself. Again, in how many ways can we represent n using n parts? Only one way, namely $1 + 1 + 1 + \cdots + 1$ (n times).

If $m > n$, then $P_n^m = 0$. Is it possible to represent n using $n + 1$ parts? Even if we use 1 in all components, we require only n parts.

Problem In how many ways can the number 8 be partitioned into four parts?

Solution We have to find P_8^4. Replacing m by 4 and n by 4 in the above recurrence relation, we obtain

$$P_8^4 = P_4^1 + P_4^2 + P_4^3 + P_4^4 = 1 + P_4^2 + P_4^3 + 1 \text{ (since } P_4^1 = P_4^4 = 1).$$

To determine P_4^2 we replace m by 2 and n by 2 in the above recurrence relation and obtain

$$P_4^2 = P_2^1 + P_2^2 = 1 + 1 = 2 \text{ (since } P_2^1 = P_2^2 = 1).$$

To determine P_4^3 we replace m by 3 and n by 1 in the above recurrence relation and obtain $P_4^3 = P_1^1 = 1$. (Note that $P_1^s = 0$ whenever $s > 1$. Is it possible to represent 1 using 2 parts?)

Putting everything together, we obtain $P_8^4 = 1 + 2 + 1 + 1 = 5$.

Various partitions of 8 using four parts are $1+1+1+5$, $1+1+2+4$, $1+1+3+3$, $1+2+2+3$ and $2+2+2+2$.

We will now try to find $P(8)$, the number of partitions of 8. We have

$$P(8) = P_8^1 + P_8^2 + P_8^3 + P_8^4 + P_8^5 + P_8^6 + P_8^7 + P_8^8. \tag{i}$$

To determine P_8^2, we replace m by 2 and n by 6 in the above recurrence relation and obtain $P_8^2 = P_6^1 + P_6^2$.

To determine P_6^2, we replace m by 2 and n by 4 in the above recurrence relation and obtain $P_6^2 = P_4^1 + P_4^2$.

To determine P_4^2, we replace m by 2 and n by 2 in the above recurrence relation and obtain $P_4^2 = P_2^1 + P_2^2 = 1 + 1 = 2$.

Hence $P_6^2 = P_4^1 + P_4^2 = 1 + 2 = 3$ and $P_8^2 = P_6^1 + P_6^2 = 1 + 3 = 4$.

To determine P_8^3, we replace m by 3 and n by 5 in the above recurrence relation and obtain $P_8^3 = P_5^1 + P_5^2 + P_5^3$.

To determine P_5^2, we replace m by 2 and n by 3 in the above recurrence relation and obtain $P_5^2 = P_3^1 + P_3^2$.

To determine P_3^2, we replace m by 2 and n by 1 in the above recurrence relation and obtain $P_3^2 = P_1^1 = 1$.

To determine P_5^3, we replace m by 3 and n by 2 in the above recurrence relation and obtain $P_5^3 = P_2^1 + P_2^2 = 1 + 1 = 2$.

Hence $P_5^2 = P_3^1 + P_3^2 = 1 + 1 = 2$ and $P_8^3 = P_5^1 + P_5^2 - P_5^3 = 1 + 2 + 2 = 5$.

To determine P_8^5, we replace m by 5 and n by 3 in the above recurrence relation and obtain $P_8^5 = P_3^1 + P_3^2 + P_3^3 = 1 + 1 + 1 = 3$.

To determine P_8^6, we replace m by 6 and n by 2 in the above recurrence relation and obtain $P_8^6 = P_2^1 + P_2^2 = 1 + 1 = 2$.

To determine P_8^7, we replace m by 7 and n by 1 in the above recurrence relation and obtain $P_8^7 = P_1^1 = 1$.

We have already found $P_8^4 = 5$. Since $P_8^1 = P_8^8 = 1$, substituting all these values in (i), we obtain $P(8) = 1 + 4 + 5 + 5 + 3 + 2 + 1 + 1 = 22$.

Can you list all the 22 partitions of 8 ?

We will now see how the method of generating functions can be used to determine $p(n)$, the number of partitions of a positive integer n. For example, consider the number 9. One partition of 9 is $1+2+2+4$. This partition says that the number 1 occurs one time, the number 2 occurs two times, the number 4 occurs one time and each of the numbers 3, 5, 6, 7, 8 and 9 occurs zero times. The numbers 1, 2 and 4 are referred to as summands. Note that a summand for 9 is a number between 1 and 9.

For any positive integer n, we are interested in determining the number of times each of the summands $1, 2, 3, \ldots, n$ occurs in each partition of n. In a partition of n, the number 1 can occur zero times (correspondingly we have 1), one time (correspondingly we have x), two times (correspondingly we have x^2) and so on. Since the number 1 can occur either zero times or one time or two times, etc., we have a term $(1+x+x^2+\cdots)$ in the generating function. Similarly, the number 2 can occur zero times (correspondingly we have 1), one time (correspondingly we have x^2), two times (correspondigly we have x^4) and so on. We thus have a term $(1 + x^2 + x^4 + \cdots)$ in the generating function. The number 3 can occur zero times (correspondingly we have 1), one time (correspondingly we have x^3), two times (correspondingly we have x^6) and so on. We thus have a term $(1 + x^3 + x^6 + \cdots)$. Proceeding like this, we have the generating function

$$P(x) = (1 + x + x^2 + \cdots)(1 + x^2 + x^4 + \cdots)$$
$$\times (1 + x^3 + x^6 + \cdots)\cdots(1 + x^n).$$

Note that n can occur zero times (correspondingly we have 1) or one time (correspondingly we have x^n). Actually, the term $1+x+x^2+\cdots$ will stop at x^n because 1 cannot occur more than n times. Similarly for the other terms. The generating function $(1+x+x^2+\cdots)(1+x^2+x^4+\cdots)(1+x^3+x^6+\cdots)\cdots(1+x^n)$ can be expressed in a more compact form as $\left(\frac{1}{(1-x)}\right)\left(\frac{1}{(1-x^2)}\right)\left(\frac{1}{(1-x^3)}\right)\cdots\left(\frac{1}{(1-x^n)}\right)$.

Consider the number 7. The corresponding generating function is

$$p(x) = (1 + x + x^2 + x^3 + x^4 + x^5 + x^6 + x^7)(1 + x^2 + x^4 + x^6)$$
$$\times (1 + x^3 + x^6)(1 + x^4)(1 + x^5)(1 + x^6)(1 + x^7).$$

We can see that the coefficient of x^7 is 15 which means the number 7 has 15 partitions.

Now consider the following partitions of 7.

$7, 1 + 6, 2 + 5, 3 + 4$ and $1 + 2 + 4$.

How can we distinguish the above partitions of 7 from other partitions of 7? We see that in each of these partitions the summands are all distinct, whereas in each of the other partitions of 7, at least two summands are the same. The above partitions are called the partitions of 7 with distinct summands. Their number denoted by $p_d(7)$ is seen to be 5.

Again consider the following partitions of 7.

$7, 1 + 1 + 5, 1 + 3 + 3, 1 + 1 + 1 + 1 + 3$ and

$1 + 1 + 1 + 1 + 1 + 1 + 1$.

How can we distinguish the above partitions of 7 from other partitions of 7? We see that in each of these partitions, all the summands are odd numbers, whereas in each of the other partitions of 7, at least one summand is an even number. The above partitions are called the partitions of 7 with odd summands. Their number denoted by $p_o(7)$ is also seen to be 5. Thus for the number 7, we have $p_d(7) = p_o(7)$.

We will now prove that for any positive integer n, $p_d(n) = p_o(n)$.

What do we do while computing $p_d(n)$? Take the number 1. In a partition of n either we do not include it as a summand (denoted by 1) or we include it only once (denoted by x). Since the summands are all distinct, we cannot include 1 more than once. We thus obtain the term $(1 + x)$. What about the number 2? Either do not include it as a summand (denoted by 1) or include it only once (denoted by x^2). We thus obtain the term $(1 + x^2)$. Proceeding like this, we obtain

$$\begin{aligned} p_d(x) &= (1 + x)(1 + x^2)(1 + x^3)\cdots \\ &= \left(\frac{(1 - x^2)}{(1 - x)}\right)\left(\frac{(1 - x^4)}{(1 - x^2)}\right)\left(\frac{(1 - x^6)}{(1 - x^3)}\right)\cdots \\ &= \left(\frac{1}{(1 - x)}\right)\left(\frac{1}{(1 - x^3)}\right)\left(\frac{1}{(1 - x^5)}\right)\left(\frac{1}{(1 - x^7)}\right)\cdots \qquad \text{(ii)} \end{aligned}$$

What do we do while computing $p_o(n)$? Take the number 1. Since 1 is an odd number, we can include it any number of times in a partition of n. We thus obtain the term $(1 + x + x^2 + x^3 + \cdots)$. Note that 1 corresponds to not including 1 at all, x corresponds to including 1 once, x^2 corresponds to including 1 two times and so on. What about the number 2? Since 2 is an even number, we do not include it even once. What about the number 3? Since 3 is an odd number, we can include it any number of times in the partition of n thus obtaining the term $(1 + x^3 + x^6 + \cdots)$. Note that 1 corresponds to not including 3 at all, x^3 corresponds to including 3 once, x^6 corresponds to including 3 two times and so on. Again we do not include the number 4 (since 4 is even) and 5 can be included any number of times giving rise to the term $(1 + x^5 + x^{10} + \cdots)$. Proceeding like this, we obtain

$$\begin{aligned} p_o(x) &= (1 + x + x^2 + x^3 + \cdots)(1 + x^3 + x^6 + \cdots) \\ &\quad \times (1 + x^5 + x^{10} + \cdots)\cdots \\ &= \left(\frac{1}{(1 - x)}\right)\left(\frac{1}{(1 - x^3)}\right)\left(\frac{1}{(1 - x^5)}\right)\left(\frac{1}{(1 - x^7)}\right)\cdots \\ &= p_d(x) \quad \text{from (ii).} \end{aligned}$$

15 Miscellaneous Applications

In this chapter we will consider some of the miscellaneous applications. In particular, we will try to answer the following very interesting questions.

1. n men go for a party. Before entering the hall they leave their overcoats in a room outside. After the party they find that the room is dark due to power failure and each one picks up an overcoat. What is the number of ways in which none will get back his own overcoat? That is, everyone gets the overcoat of another man.
2. Determine the number of positive integers less than or equal to 100 which are not divisible by 7, 9 or 11.
3. For a positive integer n, $\phi(n)$ denotes the number of positive integers which are less than n and prime to n (integers which have no common factor other than 1 with n). $\phi(n)$ is called the Euler function. For example, $\phi(8) = 4$ since the number of positive integers which are less than 8 and prime to 8 are 1, 3, 5 and 7. Find a formula for $\phi(n)$.
4. How many permutations of 1, 2, 3 and 4 are there satisfying the following?

   (i) 1 is not in the second position
   (ii) 2 is not in the third position
   (iii) 3 is not in the first or fourth position
   (iv) 4 is not in the fourth position

5. Let $X = \{1, 2, 3, \ldots, n\}$. Determine the number of subsets of X (each consisting of k numbers) such that each subset does not contain consecutive numbers. If i belongs to such a subset, neither $i - 1$ nor $i + 1$ can belong to the subset.
6. Let $X = \{1, 2, 3, \ldots, n\}$. Determine the number of subsets of X such that each subset does not contain consecutive numbers. If i belongs to such a subset, neither $i - 1$ nor $i + 1$ can belong to the subset.

7. If the numbers $1, 2, \ldots, n$ are arranged in a circle, what is the number of ways in which k of these numbers can be chosen in such a way that no two of the chosen numbers occupy consecutive positions ?
8. Determine the number of ways in which n married couples can be seated at a circular table such that men and women alternate but no husband sits next to his wife.

Before attempting to answer the above questions, we need to develop some preliminaries.

If we have three numbers, say 1, 2 and 3. They can be arranged in the following six possible ways.

1 2 3, 1 3 2, 3 2 1, 2 1 3, 2 3 1, 3 1 2.

Each of the above ways is called a permutation (arrangement) of the numbers 1, 2 and 3. Consider the last two permutations, namely 2 3 1 and 3 1 2. In these two permutations, 1 is not in the first position, 2 is not in the second position and 3 is not in the third position. These two permutations are called derangements of 1, 2 and 3.

Definition If we have n numbers, say $1, 2, 3, \ldots, n$, then a permutation of these n numbers is said to be a derangement if i does not occupy the ith position $(1 \leq i \leq n)$. In other words, a derangement of $1, 2, 3, \ldots, n$ is any arrangement of $1, 2, 3, \ldots, n$ in which 1 does not occupy first position, 2 does not occupy second position, ... n does not occupy nth position.

Consider a set S consisting of n elements. We assume that with every element $a \in S$, there is an associated weight which we will denote by $w(a)$. Let $P = \{P_1, P_2, \ldots, P_N\}$ be a set consisting of N properties. These properties depend on the particular application. Let r be an integer such that $0 \leq r \leq N$. We will define a weight $W(r)$.

When $r = 0$, $W(0)$ is simply the sum of the weights of all elements of S. In other words, $W(0) = \sum w(a)$, where the summation is taken over all elements $a \in S$.

To define $W(r)$, we consider r of the properties from $\{P_1, P_2, \ldots, P_N\}$, say $P_{i1}, P_{i2}, \ldots, P_{ir}$. We define $W(P_{i1}, P_{i2}, \ldots, P_{ir})$ to be the sum of the weights of all those elements which possess properties $P_{i1}, P_{i2}, \ldots, P_{ir}$. In other words, $W(P_{i1}, P_{i2}, \ldots, P_{ir}) = \sum w(a)$, where the summation is taken over all elements a which possess all the properties $P_{i1}, P_{i2}, \ldots, P_{ir}$. Note that the elements considered here may have properties other than $P_{i1}, P_{i2}, \ldots, P_{ir}$.

We now define $W(r) = \sum W(P_{i1}, P_{i2}, \ldots, P_{ir})$, where the summation is taken over all possible r properties chosen from $P_1, P_2, \ldots, P_N$. We now define $E(r)$

to be the sum of the weights of all those elements of S which possess exactly r properties. In particular, $E(0)$ denotes the sum of the weights of all those elements of S which possess exactly 0 properties, which is nothing but the sum of the weights of all those elements of S which do not possess any of the N properties. Note that $W(r)$ is the sum of the weights of all those elements of S which possess at least r properties. We will first try to find a formula for $E(r)$ in terms of $W(r)$.

Generalised inclusion and exclusion principle

Rather than the above principle, the following deductions of the above principle (especially Sieve formula) play vital roles in solving various problems of interest.

Result

(i) $E(0) = W(0) - W(1) + W(2) - \cdots + (-1)^N W(N)$.

(ii) If we take the weight of each element to be 1, then

$$n(0) = n - n(1) + n(2) - \cdots + (-1)^N n(N).$$

Here n denotes the number of elements in S, $n(i)$ denotes the number of elements in S which satisfy i of the N properties, and $n(0)$ denotes the number of elements in S which do not satisfy any of the N properties. Result (ii) is called the Sieve formula.

(iii) The hit polynomial $E(0) + E(1)t + E(2)t^2 + \cdots + E(N)t^N$ is given by

$$\sum_{j=0}^{N} W(j)(t-1)^j.$$

We will now try to answer Problem 1. If we number the n men as $1, 2, 3, \ldots, n$ and their overcoats also as $1, 2, 3, \ldots, n$, then the number of ways in which none gets his own overcoat is nothing but the number of derangements of $1, 2, 3, \ldots, n$. Denote this number by D_n. We will now proceed to find D_n.

We know the total number of ways in which the n numbers $1, 2, 3, \ldots, n$ can be arranged is $n!$ (There are n positions for the number 1. Once we place 1 in one of the n positions, there are only $n - 1$ positions for the number 2. Finally, there is only one position for the number n). Let P_i denote the property that in a permutation, i occupies the ith position ($i = 1, 2, 3, \ldots, n$).

Now $n(r)$ denotes the number of permutations which satisfy r properties, which is the same as the number of permutations in which r numbers are in the correct positions. These r numbers out of n numbers $1, 2, 3, \ldots, n$ can be chosen in nC_r ways. The remaining $n - r$ numbers can be arranged in $(n - r)!$ ways. It follows that $n(r) = nC_r \times (n-r)! = \frac{n!}{r!}$.

Also, $n(0)$ denotes the number of permutations which do not satisfy any of the properties. It is the number of permutations in which i does not occupy the ith position $(i = 1, 2, 3, \ldots, n)$. Hence it follows that $n(0) = D_n$. Substituting these values in the Sieve formula, we obtain

$$D_n = n! - \frac{n!}{1!} + \frac{n!}{2!} - \cdots + (-1)^n \frac{n!}{n!}$$

$$= n!\left(1 - \frac{1}{1!} + \frac{1}{2!} - \cdots + \frac{(-1)^n}{n!}\right).$$

If there are 4 men, then by substituting $n = 4$ in the above, we obtain $D_4 = 9$. If there are 5 men, then by substituting $n = 5$ in the above, we obtain $D_5 = 44$.

We will now solve Problems 2 and 3 using Sieve formula. The solutions will show how judiciously the properties are selected.

Solution to Problem 2

If n and r are two positive integers, denote by $\left[\frac{n}{r}\right]$ the largest integer not larger than $\frac{n}{r}$. For example, $\left[\frac{27}{4}\right] = 6$.

Consider a positive integer ≤ 100. Let P_1 denote the property that the positive integer is divisible by 7, P_2 denote the property that the positive integer is divisible by 9 and P_3 denote the property that the positive integer is divisible by 11. Our problem reduces to finding the number of positive integers which do not satisfy any of the above properties. The number of positive integers which satisfy P_1 is $\left[\frac{100}{7}\right] = 14$. The number of positive integers which satisfy P_2 is $\left[\frac{100}{9}\right] = 11$. The number of positive integers which satisfy P_3 is $\left[\frac{100}{11}\right] = 9$. The number of positive integers which satisfy at least one of the properties is given by $n(1) = 14 + 11 + 9 = 34$.

The number of positive integers which satisfy P_1 and P_2 is $\left[\frac{100}{63}\right] = 1$. The number of positive integers which satisfy P_2 and P_3 is $\left[\frac{100}{99}\right] = 1$. The number of positive integers which satisfy P_3 and P_1 and is $\left[\frac{100}{77}\right] = 1$. The number of positive integers which satisfy at least two of the properties is given by $n(2) = 1 + 1 + 1 = 3$.

The number of positive integers which satisfy P_1, P_2 and P_3 is $\left[\frac{100}{693}\right] = 0$. Hence the number of positive integers which satisfy at least three of the properties is given by $n(3) = 0$. Substituting these values in the Sieve formula, we see that $n(0)$, which is the number of positive integers which do not satisfy any of the properties, is given by $n(0) = 100 - 34 + 3 - 0 = 69$.

Solution to Problem 3

We know any positive integer can be expressed as a product of prime numbers and some prime numbers may even be repeating. Thus we can write n as $n = p_1^{a1} \times p_2^{a2} \times \cdots \times p_l^{al}$ ($p_1, p_2, \ldots, p_l$ are distinct primes not equal to 1 and $a_1, a_2, \ldots, a_l > 0$). Let P_i denote the property that p_i is a common factor with $n (i = 1, 2, \ldots, l)$. Then $\phi(n)$ is simply the number of elements which do not satisfy any of the properties P_i. Now $n(1)$ denotes the number of elements which satisfy at least one of the properties. It is the number of elements for which either p_1 is a factor or p_2 is a factor or p_l is a factor. The number of such elements is given by $n(1) = \sum \frac{n}{p_i}$.

Similarly,

$$n(2) = \sum \frac{n}{(p_i \times p_j)},$$

$$n(3) = \sum \frac{n}{(p_i \times p_j \times p_k)}, \qquad (i < j < k), \text{ etc.}$$

Substituting these values in Sieve formula, we obtain

$$\begin{aligned}\phi(n) &= n - n(1) + n(2) - n(3) + \cdots \\ &= n - \sum \frac{n}{p_i} + \sum_{i<j} \frac{n}{(p_i \times p_j)} - \sum_{i<j<k} \frac{n}{(p_i \times p_j \times p_k)} + \cdots \\ &= n \times \left(1 - \frac{1}{p_1}\right) \times \left(1 - \frac{1}{p_2}\right) \times \left(1 - \frac{1}{p_3}\right) \times \cdots \left(1 - \frac{1}{p_l}\right)\end{aligned}$$

Examples

1. Consider the number 13. Since 13 itself is a prime number, its representation as a product of prime numbers contains only 13. In other words,

   $$p_1 = 13, a_1 = 1.$$

   Substituting in the above formula, we obtain

   $$\phi(13) = 13 \times \left(1 - \frac{1}{13}\right) = \frac{13 \times 12}{13} = 12.$$

   Obviously, since 13 is a prime number, every number less than 13 (there are 12 such numbers) is prime to 13.

2. Now consider the number 15. We have $15 = 3 \times 5$. How many numbers less than or equal to 15 have a common factor with 3? We know this number is $\frac{15}{3} = 5$ (note that these numbers are 3, 6, 9, 12, 15). How many numbers less than or equal to 15 have a common factor with 5? We know this number is $\frac{15}{5} = 3$ (note that these numbers are 5, 10, 15). How many numbers less than or equal to 15 have a common factor with $3 \times 5 = 15$? We know this number is $\frac{15}{15} = 1$ (note that the only number is 15). Hence, by Sieve formula we have $\phi(15) = 15 - (3+5) + 1 = 15 - 8 + 1 = 8$.

Before attempting to solve Problem 4, we will develop some preliminaries.

A rook is a chessboard piece which can capture another rook which is in the same row or in the same column. Non-taking rooks are a set of rooks such that no rook in this set can be captured by other rooks in the same set. The chessboards we consider are not only the usual ones but also those which can have holes. They can even have disconnected components. Let r_k denote the number of ways in which k non-taking rooks can be placed on a chessboard.

We can associate with the sequence $(r_0, r_1, r_2, \ldots)$ the series $R_k(C) = r_0 + r_1 t + r_2 t^2 + \cdots$, where C stands for the chessboard. This polynomial is called the rook polynomial of the chessboard C.

In how many ways two non-taking rooks can be placed on the usual chessboard?

We know the usual chessboard has 8 rows and 8 columns. Since the rooks are non-taking, they have to occupy different rows. The number of ways in which 2 rows can be selected out of 8 rows is ${}^8C_2$. In one of the selected rows, one rook can be placed in one of the 8 columns. Thus there are 8 choices for this rook. In the other selected row, the other rook can be placed only in 7 possible ways because the other rook cannot be placed in the same column where the first rook is placed. Thus the total number of ways is ${}^8C_2 \times 8 \times 7$.

We will now try to find the rook polynomial for some chessboards. Consider the chessboard given below.

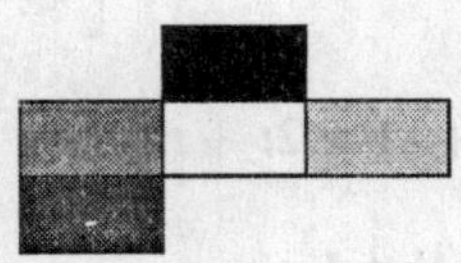

If we have only one rook, obviously it can be placed in one of the 5 cells. Hence $r_1 = 5$. It can be verified that two non-taking rooks can be placed in 5 ways and three non-taking rooks can be placed in only 1 way. Also, it can be seen that the number of ways in which four or more than four non-taking rooks can be placed is 0. Hence $r_2 = 5, r_3 = 1$ and $r_k = 0 (k \geq 4)$. By convention, $r_0 = 1$. Thus the rook polynomial becomes $1 + 5t + 5t^2 + t^3$.

We will now develop a recurrence relation to determine the rook polynomial of an arbitrary chessboard C. Suppose we mark one cell of the chessboard C as a distinguished cell d. Let C_{dd} denote the chessboard obtained by deleting the row and column in which the distinguished cell d appears. Let C_d' denote the chessboard obtained by deleting the cell d. Consider the problem of placing k non-taking rooks in C. The following possibilities occur.

(i) One of the k rooks occupies cell d. Since the rooks are non-taking, none of the remaining $k - 1$ rooks can occupy a cell which is in the same row or same column as cell d. In other words, the remaining $k - 1$ non-taking rooks are to be placed in the chessboard C_{dd}.

(ii) None of the k rooks occupies cell d. In this case the problem reduces to placing k non-taking rooks in the chessboard C'_d.

It follows that $r_k(C) = r_{k-1}(C_{dd}) + r_k(C'_d)$ and hence

$$R(t, C) = tR(t, C_{dd}) + R(t, C'_d).$$

The following examples will show that the rook polynomial of a chessboard is independent of the cell which is marked as a distinguished cell. In other words, whichever cell is marked as a distinguished cell, the rook polynomial of a chessboard will remain the same.

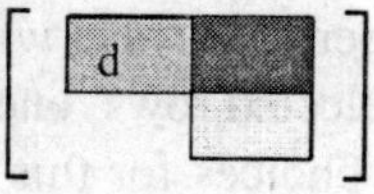

$$[\text{d}] = t\,[\ ] + [\ ]$$

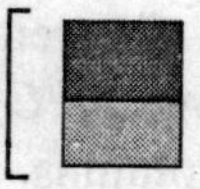

$$= t(1 + t) + (1 + 2t)$$

$$= 1 + 3t + t^2$$

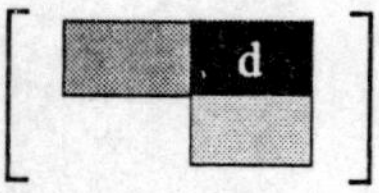

$$[\text{d}] = t\,[\text{Null diagram}] + [\ ]$$

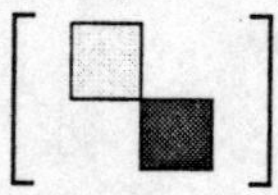

$$= t.1 + (1 + 2t + t^2)$$

$$= 1 + 3t + t^2$$

Suppose a chessboard C can be decomposed into two chessboards C_1 and C_2 such that if some cells from ith row of C are present in C_1, then no cells from the ith row of C are present in C_2 and vice versa. Similarly for the columns. Then the movements of rooks in C_1 do not affect the movements of rooks in

C_2 and vice versa. In other words, the movements of rooks in C_1 and C_2 are independent. It follows that $R(t, C) = R(t, C_1) \times R(t, C_2)$.

The following examples will illustrate how this fact is made use of in determining the rook polynomial of a chessboard C.

$$= t[\;] + [\;]$$

$$= t(1+t) + [\;] \times [\;]$$

$$= t(1+t) + (1+t)\left[t(\text{Null diagram}) + \;\right]$$

$$= t(1+t) + (1+t)[t.1 + 1 + 2t + t^2]$$

$$= 1 + 5t + 5t^2 + t^3$$

We now try to solve Problem 4 using the method of rook polynomials. Since there are 4 numbers, we consider a chessboard which has 4 rows and 4 columns. 1 should not be in the second position implies that we place a cross at the cell which is in the first row and the second column. Similarly, we place crosses at the other cells and obtain the following chessboard.

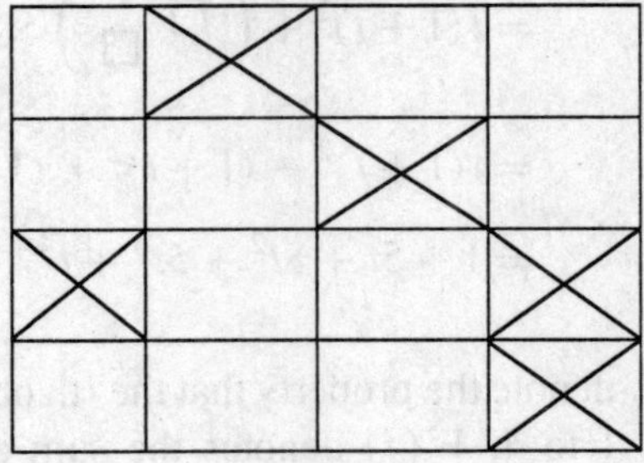

Now the rooks can be placed only in those cells which are not crossed. Consider a permutation say

$$\begin{matrix} 1 & 2 & 3 & 4 \\ j_1 & j_2 & j_3 & j_4 \end{matrix}$$

This corresponds to placing a rook in the cell which is at row 1 and column j_1 (obviously j_1 cannot be 2), a rook in the cell which is at row 2 and column j_2

(obviously j_2 cannot be 3), a rook in the cell which is at row 3 and column j_3 (obviously j_3 cannot be 1 or 4), and finally a rook in the cell which is at row 4 and column j_4 (obviously j_4 cannot be 4). If we place a rook in the first row, no other rook can be placed in the first row. If we do this, 1 will have two images which cannot happen in a function. If we place a rook in the cell which is at row 1 and column 3 (in this case, $j_1 = 3$), then no other rook can be placed in column 3. Suppose we place a rook in the cell which is at row 4 and column 3 (in this case, $j_4 = 3$). Then both 1 and 4 will have the same image. This again cannot happen since a permutation is a one–one function.

The above discussions simply mean that the answer to the problem is the number of ways in which 4 non-taking rooks can be placed on the chessboard C obtained by removing the cells which are crossed. This chessboard is shown alongside.

Instead of finding the rook polynomial of the chessboard C which may be difficult, we will determine the rook polynomial of the chessboard of forbidden positions.

$$[\ \text{d}\ ] = t[\ \ ] + [\ \ ]$$

$$= t(1+t)^2 + \left(\square\ _{\square}\right) \times \left(\square\!\square\right)$$

$$= t(1+t)^2 + (1+t)^2 \times (1+t)^2$$

$$= 1 + 5t + 8t^2 + 5t^3 + t^4 \qquad \text{(i)}$$

For each $i = 1$ to 4, let P_i denote the property that the ith object is in a forbidden position. For each $j = 1$ to 4, $W(j)$ denotes the sum of the weights of all those elements which satisfy at least j properties. If we take the weight of each element to be 1, then $W(j)$ is simply the number of elements which satisfy at least j properties. Since each property says that one object is in a forbidden position, to find $W(j)$ we have to find the number of ways in which j objects can be in forbidden positions. This is the same as determining the number of ways in which j non-taking rooks can be placed on the chessboard C which is r_j (note that r_js are given by the coefficients of (i)). Since the remaining $n - j$ objects can be placed in $(n - j)!$ ways, it follows that $W(j) = r_j \times (n - j)!$ Since $n = 4$, we obtain the hit polynomial

$$
\begin{aligned}
&E(0) + E(1)t + E(2)t^2 + E(3)t^3 + E(4)t^4 \\
&= r_0 \times n! \times (t-1)^0 + r_1 \times (n-1)! \times (t-1)^1 + r_2 \times (n-2)! \\
&\quad \times (t-1)^2 + r_3 \times (n-3)! \times (t-1)^3 + r_4 \times (n-4) \times (t-1)^4. \\
&= 4! + 5 \times 3! \times (t-1) + 8 \times 2! \times (t-1)^2 + 5 \times 1! \times (t-1)^3 \\
&\quad + (t-1)^4 \\
&= 24 + 30 \times (t-1) + 16 \times (t-1)^2 + 5 \times (t-1)^3 + (t-1)^4 \\
&= 6 + 9t + 7t^2 + t^3 + t^4 \text{ (after simplification).}
\end{aligned}
$$

Equating the coefficients of t, t^2, t^3, t^4 and the constant terms, we obtain

$E(0) = 6$, which gives the number of ways in which none of the integers is in a forbidden position.

$E(1) = 9$, which gives the number of ways in which exactly one integer is in a forbidden position.

$E(2) = 7$, which gives the number of ways in which exactly two integers are in forbidden positions.

$E(3) = 1$, which gives the number of ways in which exactly three integers are in forbidden positions.

$E(4) = 1$, which gives the number of ways in which exactly four integers are in forbidden positions.

We will now consider Problem 5.

Let $f(n, k)$ denote the number of subsets of X (each consisting of k numbers) such that no subset contains consecutive numbers. Before solving the problem, we will understand the problem by means of an example. If $X = \{1, 2, 3, 4\}$ and $k = 2$, then the required subsets are $\{1, 3\}$, $\{1, 4\}$ and $\{2, 4\}$. Hence the required number is 3. Note that we have not considered the subsets $\{1, 2\}$, $\{2, 3\}$ and $\{3, 4\}$ since all of them contain consecutive numbers. It follows that $f(4, 2) = 3$. Note that $(n - k + 1)C_k = (4 - 2 + 1)C_2 = 3C_2 = 3$.

Solution to Problem 5

Consider a subset S of X (consisting of k numbers) which does not contain consecutive numbers. We can associate with S a sequence $a_1a_2a_3 \ldots a_n$ as follows.

$$
\begin{aligned}
a_i &= 1 \text{ if } i \in S \\
&= 0 \text{ if } i \notin S
\end{aligned}
$$

We see that this is a sequence of 0s and 1s whose length is n. Since S does not contain consecutive numbers, the sequence cannot have consecutive 1s. The association $S \to a_1a_2a_3\ldots a_n$ can be seen to be a one–one correspondence between the collection of all subsets of X (each subset consisting of k numbers and no subset containing consecutive numbers) and the collection of all sequences of 0s and 1s of length n such that no sequence has consecutive 1s in it. Thus it suffices to count the number of such sequences.

Since each subset contains k numbers, it follows that each of the sequences of length n has k 1s and the remaining $n-k$ numbers are 0s. We will first write these $n-k$ 0s as follows.

$$0 \quad 0 \quad 0 \cdots 0$$

We have to now insert k 1s. How many positions are there for their insertions? 1 can be inserted at the extreme left or at the extreme right. These give two positions. 1 can be inserted (maximum once since no two 1s are consecutive) between any two 0s. Since the number of 0s is $n-k$, the number of such positions is $n-k-1$. Hence the total number of positions where 1 can be inserted is $n-k-1+2 = n-k+1$. Out of these $n-k+1$ positions, we have to select k positions (since we want only k 1s) and this can be done in $(n-k+1)C_k$ ways. Hence it follows that $f(n, k) = (n-k+1)C_k$.

Now consider Problem 6. In Problem 5 we considered only those subsets of X which consist of k numbers, whereas in this problem we do not impose any restrictions on the number of elements in each subset. In other words, we consider all possible subsets of X (we know its number is 2^n) and remove from this collection all those subsets which contain consecutive numbers. Since the number now depends only on n (and not on k), we will denote the required number by $F(n)$. The required class of subsets can now be divided into two classes.

1. One class consists of all those subsets of X (none of which contains consecutive numbers) which contain n. Then, by our requirement these subsets do not contain $n-1$ (if they do contain $n-1$, then these subsets contain adjacent numbers namely n and $n-1$). Removing the two numbers n and $n-1$, we have a set consisting of $n-2$ numbers from 1 to $n-2$ and the problem reduces to finding the number of subsets of this set none of which contains consecutive numbers. We know this number is $F(n-2)$.
2. Another class consists of all those subsets of X (none of which contains consecutive numbers) which do not contain n. Then all these subsets are in fact the subsets of $\{1, 2, \ldots, n-1\}$. We know the number of such subsets is $F(n-1)$.

Thus the required number is given by the following recurrence relation.

$$F(n) = F(n-1) + F(n-2).$$

We can identify the above as a Fibonacci sequence.

Now consider Problem 7. Note the difference between this problem and Problem 5. In Problem 5, subsets of $X = \{1, 2, \ldots, n\}$ can contain both 1 and n. They cannot contain 1 and 2, 2 and 3, ..., $n-1$ and n. In this problem, in addition to the above pairs, a subset also cannot contain both 1 and n. Denote the number of such subsets by $g(n, k)$. For the above example, $g(n, k) = 2$ (note that we have to consider only the pairs (1, 3) and (2, 4). We should not consider the pair (1, 4). We also note that

$$\frac{n}{n-k} \times (n-k)C_k = \frac{4}{2} \times 2C_2 = 2.$$

The required subsets (each containing k numbers but no consecutive numbers including 1 and n) can be divided into the following two classes.

1. A class which consists of all those subsets which contain n. Then these subsets do not contain $n-1$ (since no subset contains consecutive numbers) and 1 (by the requirement of this problem). Eliminating these three numbers (n, $n-1$ and 1), we have $n-3$ numbers and we have to find the number of subsets of the set consisting of $n-3$ numbers (each subset containing $k-1$ numbers since each subset already contains n) such that each subset does not contain consecutive numbers. By Problem 5, this number is $f(n-3, k-1)$.
2. A class which consists of all those subsets which do not contain n. Then each subset is in fact a subset of $\{1, 2, \ldots, n-1\}$, contains k numbers and does not contain consecutive numbers. Again by Problem 5, the number of such subsets is $f(n-1, k)$. We thus obtain

$$g(n, k) = f(n-3, k-1) + f(n-1, k).$$

Substituting the values of f from Problem 5 and simplifying, we obtain

$$g(n, k) = \frac{n}{n-k} \times (n-k)C_k.$$

Now consider Problem 8. First we will see in how many ways n wives can be seated. Since there are $2n$ chairs (n for men and n for women), one woman can choose any of the $2n$ chairs. Hence she has $2n$ choices. Since men and women occupy alternate chairs, once a wife has occupied a chair, n chairs in which n husbands can sit and $n-1$ chairs in which the remaining $n-1$ wives can sit are fixed. Hence another woman can occupy one of the $n-1$ chairs (which are reserved for wives) which can be done in $n-1$ ways. A third woman can occupy one of the $n-2$ chairs and finally the last woman has only one choice. It follows that the number of ways in which n wives can be seated is $2n \times (n-1) \times (n-2) \times \cdots \times 1 = 2 \times n!$. Assume that the wives are seated in the order $W_1, W_2, \ldots, W_n$ (either clockwise or anticlockwise) and their husbands are labelled $H_1, H_2, \ldots, H_n$ respectively. Label the vacant seats as $1, 2, 3, \ldots, n$ with label 1 on a vacant chair between the chairs occupied by W_1 and W_n. Since a husband should not occupy a chair next to his wife, husband H_i should occupy a chair which does not lie between the chairs occupied by W_i and W_{i+1} and also between the chairs occupied by W_{i-1} and $W_i (i = 2, 3, \ldots, n-1)$. Also, H_1 should occupy a chair which does not lie between the chairs occupied by W_1 and W_2 and by W_1 and W_n and H_n should occupy a chair which does not lie between the chairs occupied by W_{n-1} and Wn and by W_n and W_1. Suppose this arrangement of seating of n husbands can be carried out in U_n ways. Then the answer to our problem will be $2 \times n! \times U_n$.

We now come to the problem of evaluating U_n. Let P_i denote the property that the permutation has i in the ith place and Q_i denote the property that the permutation has i in the $(i+1)$th place (mod n). Then the $2n$ properties $P_1, P_2, \ldots, P_n, Q_1, Q_2, \ldots, Q_n$ together are incompatible. If P_i holds, then Q_i does not hold. If i is in the ith place, how can i be in $(i+1)$th place (mod n)? We select k of these properties; we want to find the number of permutations which satisfy each of these k properties. If the k properties are not compatible, then there can be no permutation which satisfies each of the k properties. If the k properties are compatible, then we know the positions of k numbers. If P_i holds, then i is in the ith place. If Q_j holds, then j is in the $(j+1)$th place (mod n). Remaining $n-k$ numbers can be arranged in $(n-k)!$ ways. It follows that the number is $(n-k)!$ if k properties are compatible and 0 otherwise.

Now the question is, how to select k compatible properties out of the $2n$ properties $P_1, P_2, \ldots, P_n, Q_1, Q_2, \ldots, Q_n$. Let v_k denote the number of ways in which this can be done. If P_i is selected, then Q_i should not be selected $(1 \le i \le n)$. If Q_i is selected, then P_{i+1} should not be selected $(1 \le i \le n-1)$. Arrange the properties in a circle in the order $P_1, Q_1, P_2, Q_2, \ldots, P_n, Q_n$. Then we can see that only the adjacent pairs are not compatible. Hence the problem reduces to selecting subsets of $\{P_1, Q_1, P_2, Q_2, \ldots, P_n, Q_n\}$, each consisting of k elements such that none of these subsets contains adjacent elements as also both P_1 and Q_n. From Problem 7 we know that the number of such subsets is $g(2n, k)$, proving that

$$v_k = g(2n, k) = \frac{2n}{2n-k} \times (2n-k)C_k.$$

Using Sieve formula and substituting the values of $v_1, v_2, v_3, \ldots, v_n$, we obtain

$$U_n = \text{number of permutations which satisfy none of the } 2n \text{ properties}$$

$$= n! - v_1 \times (n-1)! + v_2 \times (n-2)! - v_3 \times (n-3)! + \cdots + (-1)^n v_n \times (n-n)!$$

$$= n! - \frac{2n}{(2n-1)} \times (2n-1)C_1 \times (n-1)! + \frac{2n}{(2n-2)} \times (2n-2)C_2 \times (n-2)! - \frac{2n}{(2n-3)} \times (2n-3)C_3 \times (n-3)! + \cdots + (-1)^n \frac{2n}{n} \times n\, C_n \times 0!$$

If there are four couples, then $U_4 = 2$ and the number of ways $= 96$.

If there are five couples, then $U_5 = 12$ and the number of ways $= 2880$.

If there are six couples, then $U_6 = 44$ and the number of ways $= 63360$.

Index